Engineering Calculation Methods for Turbulent Flow

Engineering Calculation Methods for Turbulent Flow

PETER BRADSHAW
Department of Aeronautics
Imperial College of Science and Technology
London

TUNCER CEBECI
Mechanical Engineering Department
California State University
and
Research Aerodynamics Subdivision
Douglas Aircraft Company
Long Beach
California

JAMES H. WHITELAW
Department of Mechanical Engineering
Imperial College of Science and Technology
London

Academic Press 1981

Copyright © 1981

Preface

Many calculation methods have been developed for turbulent flows and they provide useful information over limited ranges of boundary conditions. Correlation equations, based on mathematical expressions with coefficients determined by statistical methods and on the basis of measurements, allow interpolation within the range of the measurements and are widely used, for example in the process industries. Methods of solving integral equations representing conservation of momentum and scalar quantities have been described, for example by Walz (1966), and found favour for boundary-layer flows particularly up to the early 1970s; of course they also require information from experiments. Correlation equations have been and are applied to a wide range of flows extending from boundary layers to multiphase separated flows but the range of validity of any one correlation equation is small. Integral methods have been applied to a more limited range of flows and mainly to two-dimensional boundary layers in the absence of severe pressure gradients where they allow extrapolation with greater confidence than correlations.

The development of integral methods has been overtaken, in recent years, by methods based on the solution of partial differential equations which are the subject of this book. They can be applied to a wide range of flows which include, for example, three-dimensional boundary layers on wings as shown by Cebeci *et al.* (1977) and the complex, three dimensional combusting separated flows in gas-turbine combustors as reported by Jones and Priddin (1978). Differential methods are undoubtedly more generally applicable than alternatives but also require an input from experiments and can be less precise, over small ranges of flow, than correlation equations. There is no alternative to the use of differential methods for the range of flows relevant to engineering practice and to the more limited range familiar to the authors and considered here.

The following fifteen chapters stem from lecture courses delivered at the California State University in Long Beach and at the Imperial College in London, over the past two years. These courses were presented in three-day periods, and provided engineers in industry, government laboratories and universities with knowledge of calculation methods based on finite-difference solutions of relevant conservation equations in differential form. It is impossible to communicate, either in a short course or in a book of reasonable size, all that is known of engineering calculation methods and a choice was therefore necessary. As a result, the present emphasis is on the conservation equations and physical assumptions necessary to characterize turbulent flow (Chapters 1 to 4) and on numerical procedures for the calculation of the flow around airfoils and wings. This choice implies that most of the discussion of numerical methods is presented in the context of boundary-layer equations (Chapters 5 to 8) and that the numerical procedure of Chapter 9 is of this type.

More advanced topics, including more complex flows, are considered in Chapters 10 to 15.

It should be appreciated that the material of the first four chapters is common to all numerical methods for flow and convective heat-transfer problems, including that of Chapters 5 to 9 and those discussed in the remainder of the book. Similarly, some of the viscid–inviscid interactions of Chapter 11, the stability and transition of Chapter 12, the wing of Chapter 13 and some aspects of turbomachinery and combusting flows can make use of the procedure of Chapters 5 to 9; others can make use of methods appropriate to the solution of elliptic equations such as that of Chapter 10.

The present contribution can usefully be viewed in relation to alternative books. That of Cebeci and Bradshaw (1977), for example, is concerned with momentum transfer in thin shear layers and contains some of the material of Chapters 5, 8 and 9 in particular; it is, however, less applied and addresses itself to a narrower range of applications. The earlier paper by Harlow and Nakayama (1967) and the book by Patankar and Spalding (1970) are also less engineering oriented; the former is general in its approach to numerical methods and the latter is concerned with one boundary-layer procedure. The following four chapters, concerned with equations and turbulence models, are shorter but more up-to-date than the material in Launder and Spalding (1972). The emphasis on a single type of calculation method, in the present text, may be examined in the context of the others discussed in subsequent chapters; all are closely related to flow configurations of practical relevance.

The format of the present book reflects its origins. The use of panels containing the more important facts and topics for discussion has been retained. The discussion under each panel expands the topics of the panel. It is hoped that readers will find this approach convenient and economical of their time. The references are collected at the end of the book and include a high proportion of the most recent books and papers. Standard "bookwork" like the logarithmic velocity profile law analysis (panel 1.11) is given only in outline: readers who wish for details can find them in the references cited.

The notation, for most purposes, makes use of rectangular Cartesian coordinates x, y, z and the corresponding instantaneous velocity components, $\hat{U}, \hat{V}, \hat{W}$. The time–mean velocity components are U, V, W and the time-dependent fluctuations about these means u, v, w so that $\hat{U} \equiv U + u$, etc. Other mean values, such as the Reynolds normal stress $\overline{u^2}$, are denoted by overbars. Scalar properties, for example, pressure P, temperature T and density ρ, are represented in the same way except where ambiguity can occur. Thus a prime is added to designate the fluctuating component of some properties, for example $\hat{T} = T + T'$ and $\hat{\rho} = \rho + \rho'$. Occasionally tensor notation is used, with x_i standing for x_1, x_2 or x_3 and U_i for U_1, U_2 or U_3. Note, for example, the repeated-suffix summation convention

$$\frac{\partial U_i}{\partial x_i} = \frac{\partial U_1}{\partial x_1} + \frac{\partial U_2}{\partial x_2} + \frac{\partial U_3}{\partial x_3}$$

Preface

In at least one respect, the terminology of the book is not entirely consistent. The phase "boundary-layer equations" is commonly used to describe the reduced form of the Navier–Stokes equations which applies to the shear layer in the immediate proximity of a surface. The same equations apply to free flows such on jets, wakes and mixing regions and, in recent years, the phrase has been used for these applications where no surface exists. As a consequence, "thin-shear-layer equations" is perhaps a better term and it has been introduced in the following pages, especially when free flows, described by the two-dimensional, parabolic form of the Navier–Stokes equations, are under consideration.

We are grateful to our secretaries, Sally Chambers, Jan Davies, Roslee Fairhurst and especially Nancy O'Barr, for their considerable help in the preparation of these notes; to our research sponsors and notably the Science Research Council, Morton Cooper, David Siegel and Robert Whitehead of the Office of Naval Research and William Volz of the Naval Air Systems Command, for supporting our studies of turbulent flow; to our colleagues, research assistants and students for their contributions to these studies. Finally, we wish to acknowledge the tireless help of Dr Hillar Unt in the organization of the courses at California State University.

November 1980 PETER BRADSHAW
Palm Springs TUNCER CEBECI
California JAMES H. WHITELAW

Contents

Preface — v

Nomenclature — xi

1. Introduction — 1
2. Conservation Equations and Boundary Conditions — 22
3. Turbulence Models based on Eddy Viscosity Hypotheses — 37
4. Stress Equation Modelling — 58
5. Introduction to Numerical Methods for the Solution of the Thin Shear-layer Equations — 77
6. Solution of Two-dimensional External Boundary-layer Problems — 101
7. Inverse Boundary-layer Problems — 123
8. Unsteady, Two-dimensional and Steady, Three-dimensional Flows — 138
9. Computer Program for Unsteady Two-dimensional Boundary Layers — 159
10. Recirculating Flows — 192
11. Viscous–Inviscid Interactions and Corner Flows — 219
12. Stability and Transition — 235
13. Wings — 253
14. Turbomachinery — 273
15. Combustion — 290

References — 310

Subject Index — 327

Nomenclature

Numbers indicate pages on which quantities are defined or first used. Occasional notation is defined where used.

A or A^+	Van Driest damping constants (13, 43)
b	$\nu + \nu_t$
C	logarithmic law constant (14)
c_μ	eddy viscosity factor (39)
D/Dt	$\equiv U\dfrac{\partial}{\partial x} + V\dfrac{\partial}{\partial y} + W\dfrac{\partial}{\partial z}$ mean substantial derivative (3)
d/dt	$\equiv \dfrac{\partial}{\partial t} + \hat{U}\dfrac{\partial}{\partial x} + \hat{V}\dfrac{\partial}{\partial y} + \hat{W}\dfrac{\partial}{\partial z}$ instantaneous substantial derivative (2)
H	total enthalpy (24); shape factor δ^*/θ (113)
h	half-height of duct (125)
k	roughness height (14), conductivity (24), turbulent kinetic energy (29)
L	eddy length scale (general definition 5, specific definition $(-\overline{uv})^{3/2}/\varepsilon$, 70)
L_x	integral length scale (6)
L_ε	dissipation length scale (6)
l	mixing length (6)
l_1	mixing length (39)
M	Mach number (8; also Ma), molar concentration (300)
m	pressure gradient parameter $(x/U_e)\,dU_e/dx$ (108)
P	turbulent energy production rate (38)
Pr	Prandtl number $\mu c_p/k$
R	gas constant (306) or Reynolds number
Re	Reynolds number (length scale indicated by subscript)
r	radius (especially in axisymmetric coordinates)
S	source term in conservation equations (24)
U_τ	$(\tau_w/\rho)^{1/2}$ (14)
α	incidence (149), x-wise wave number (237)
β	z-wise wave number (237)
Γ	circulation (15), molecular or turbulent diffusivity of scalar (34)
γ	intermittency (in outer layer or transition region)
δ	boundary-layer thickness
δ_{ij}	Kronecker delta: equal to one if $i = j$, else zero
δ^*	displacement thickness (99)
ε	turbulent kinetic energy dissipation rate (6)
η	dimensionless normal coordinate (108)
θ	momentum thickness (18), circumferential coordinate

κ	logarithmic law constant (14)
λ	molecular diffusivity (33)
ν	kinematic viscosity; $\mu = \nu\rho$
ν_t	eddy (turbulent) kinematic viscosity (6); $\mu_t = \nu_t\rho$
Π	boundary-layer "wake" parameter (18)
σ_ϕ	turbulent Prandtl/Schmidt number (54)
τ	shear stress (13); occasionally, (shear stress)/density
ψ	stream function
Ω	vorticity (15)
ω	vorticity or angular velocity

Subscripts

b	bulk (cross-sectional average)
c	chord (in Re_c)
fu	fuel
i, j	tensor subscripts: terms in which a subscript appears twice are summed over all values (1, 2, 3) of that subscript
0	initial or reference conditions
ox	oxidant
w	wall (solid surface)

Superscripts

$\hat{\ }$	instantaneous value
$\bar{\ }$	mean value

Eddy viscosity distribution for CS method; page 43
Constants in k–ε method; page 48
Finite difference notation; page 83

Introduction

1.1 Purposes and outline of chapter

- To introduce the conservation equations for momentum and Reynolds stress, to illustrate the meanings of different terms and to comment on the strategy of calculation methods
- To present a general introduction to the physics of turbulence, especially in shear layers

In the first few panels the derivation of the Reynolds-averaged mean-motion equations (containing the Reynolds stresses as unknowns) and of the Reynolds stress transport equations is outlined, using control-volume analysis. Mathematical details will appear in later chapters. These are *exact* equations but, being time averages of the Navier–Stokes equations, they do not contain enough information about the turbulence to form a closed, soluble set. The process of replacing the exact but insoluble equations by approximate but soluble ones is called "modelling".

Most turbulent flows of engineering importance are shear layers in which the most important Reynolds stress gradient is that of the shear stress. Some shear layers are so thin that other Reynolds stress gradients, and the change of pressure through the thickness of the layer, are negligible. Wall shear layers (boundary layers, duct flows, etc) are dominated by the inner layer, to which similarity rules apply to a good approximation.

The basic mechanism of turbulence (and of transition from laminar flow) is the stretching and distortion of elementary vortex lines by each other and by the mean flow; energy is extracted from the mean flow and cascaded down to the smallest eddies. ("Eddies" are packages of vortex lines.) Shear layers are frequently affected by externally imposed distortions or by interaction with another turbulence field.

1.2 Control-volume analysis

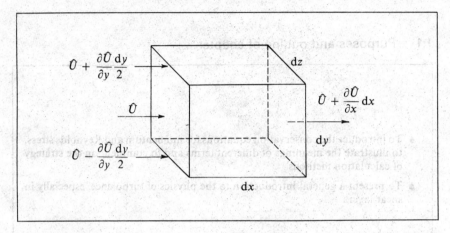

The easiest way to derive and discuss the equations is control-volume analysis. The net rate of increase of, say, the x-component momentum of the fluid in the control volume is $(\partial(\rho\hat{U})/\partial t)\,dx\,dy\,dz$ where $\hat{U}$ is the instantaneous velocity, $U + u$. The net rate of transport of x-component momentum out of the control volume is

$$[\partial(\rho\hat{U}\hat{U})/\partial x + \partial(\rho\hat{U}\hat{V})/\partial y + \partial(\rho\hat{U}\hat{W})/\partial z)]\,dx\,dy\,dz.$$

The sum of these two, $(d(\rho\hat{U})/dt)\,dx\,dy\,dz$, equals the rate at which the x-component momentum of the fluid in the control volume is increased by gradients of pressure or of viscous stresses. If the time mean is taken the first term disappears, the pressure and viscous terms are the same as in steady flow, but because, for example,

$$\rho\overline{\hat{U}\hat{V}} \equiv \rho\overline{(U + u)(V + v)} = \rho UV + \rho\overline{uv},$$

the momentum transport terms contain extra mean parts depending on the velocity fluctuations. Transferred to the right-hand side and regarded as extra apparent stresses, $-\overline{\rho u^2}$ is the normal stress in the x direction and $-\overline{\rho uv}$ and $-\overline{\rho uw}$ are shear stresses acting in the x direction on the faces perpendicular to the y and z directions respectively. Similar groups of three extra stresses appear in the y- and z-component momentum equations. Collectively they are the Reynolds stresses $-\overline{\rho u_i u_j}$.

Introduction

1.3 Transport equations for Reynolds stresses

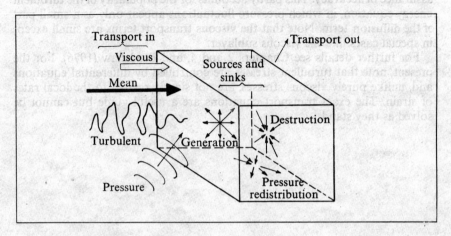

By multiplying the $d\hat{U}_i/dt$ equations by u_j, adding u_i times the $d\hat{U}_j/dt$ equation and taking the time mean, we obtain an equation for $D\overline{u_i u_j}/Dt$. The $\overline{u_i u_j}$ are the unknown turbulence quantities in the mean momentum-conservation equations. The $D\overline{u_i u_j}/Dt$ equations contain further unknowns and so on—the only exact closed set are the original time-dependent Navier Stokes equations. Like the momentum-conservation equations, the $D\overline{u_i u_j}/Dt$ equations contain terms expressing transport by the mean flow ("convection"), mean transport by the fluctuations, and terms representing generation or destruction within the control volume. In general the $\overline{u_i u_j}$ are not conserved in the thermodynamic sense but the scalar $\frac{1}{2}\overline{u_i^2}$, or $(\frac{1}{2})\overline{(u^2 + v^2 + w^2)}$, is. It is the turbulent kinetic energy per unit mass. The generation term in the turbulent energy equation is the sum of all the products of Reynolds stresses and the corresponding rates of mean strain: stress × rate of strain = rate of doing work = rate of "production" of energy. The destruction term is the sum of mean products of fluctuating viscous stresses and fluctuating strain rates; that is, the mean rate at which the turbulence does work against viscous stresses and "dissipates" turbulent energy into thermal internal energy. The mean-transport term $D\frac{1}{2}\overline{u_i^2}/Dt$ is called "advection" and the turbulent-transport term "diffusion". In most cases, transport terms are fairly small and sometimes negligible.

The "pressure redistribution" terms in the digram almost always tend to make the turbulence more nearly isotropic—they tend to reduce shear stresses ($-\rho\overline{u_i u_j}$ with $i \neq j$) and to make the three normal stresses, $-\rho\overline{u_1^2}$, $-\rho\overline{u_2^2}$ and $-\rho\overline{u_3^2}$ more nearly equal (without changing the turbulent energy $\frac{1}{2}\overline{u_i^2}$). The "pressure redistribution" term in the $\overline{u_i u_j}$ equation is the mean product of the pressure fluctuation and the fluctuating rate of strain in the $x_i x_j$ plane, and so is called the "pressure-strain" term: see panels 2.7, 4.7.

We cannot measure pressure fluctuations within the fluid with any assurance of accuracy. This partly accounts for the popularity of the turbulent energy equation, in which pressure fluctuations appear only as a small part of the diffusion term. Note that the viscous transport terms are small except in special cases like the viscous sublayer.

For further details see Chapters 2 and 4, and Bradshaw (1976). For the present, note that turbulent stresses are controlled by differential equations and, unlike purely viscous stresses, are not simply related to the local rates of strain. The exact transport equations are a useful guide but cannot be solved as they stand.

Introduction

1.4 Closure of the transport equations ("modelling")

$$\frac{DU_i}{Dt} = \frac{1}{\rho}\frac{\partial}{\partial x_j}(-\overline{\rho u_i u_j}) + \ldots$$

$$\frac{D}{Dt}(\overline{u_i u_j}) = -\frac{\partial}{\partial x_l}(\overline{u_i u_j u_l}) + \ldots$$

(all terms have dimensions velocity^3/length)

$$\frac{D}{Dt}V^m L^n = mV^{m-1}L^n \frac{DV}{Dt} + nV^m L^{n-1}\frac{DL}{Dt} \ldots$$

The mean-momentum transport equations for U_i contain second-order products, the Reynolds stresses; the Reynolds stress transport equation contain third-order products (and other unknown turbulence quantities), and so on. An infinite set of time-mean equations would be needed to reproduce all the information in the instantaneous Navier Stokes equations. To calculate the U_i, the set of equations must be truncated by using experimental data or inspired guesses. *Any* calculation method for turbulent flow should be explicable, at least qualitatively, as such a truncation: whether it uses the Reynolds stress transport equations explicitly or not, a calculation method should not contradict them.

Most calculation methods "model" one or more of the Reynolds stress transport equations, by expressing the unknown turbulence quantities on the right-hand side as empirical functions of the $\overline{u_i u_j}$, the mean strain rate, etc. The dimensions of $D\overline{u_i u_j}/Dt$ are (velocity)^3/(length) and a length scale, L say, must be introduced into at least one of the terms. Given a velocity scale, V say, (perhaps $(\overline{u_i^2})^{1/2}$), any quantity like $V^m L^n$ can be used for this, and a length scale recovered explicitly by using the velocity scale.

For a general review see Reynolds (1976) or Rodi (1978).

1.5 Choice of length scale

$l \equiv (-\overline{uv})^{1/2}/(\partial U/\partial y)$ mixing length

$v_t = [VL] \equiv -\overline{uv}/(\partial U/\partial y)$ eddy viscosity

$L_x \equiv \int_0^\infty \overline{u(\mathbf{x})u(\mathbf{x}+\mathbf{r}_1)}\, dr_1/\overline{u^2}(\mathbf{x})$ integral scale

$L_\varepsilon \equiv \dfrac{(\overline{u_i^2}/2)^{3/2}}{\varepsilon}$ where $\varepsilon = \dfrac{v}{2}\left(\dfrac{\partial u_i}{\partial x_j} + \dfrac{\partial u_j}{\partial x_i}\right)^2$

There are 26 more integral scales like L_x

In the mixing length formula (which can be explained as a model of the shear-stress transport equation neglecting *all* the transport terms) empirical information is inserted entirely via the mixing length itself. Only if the transport terms really are small (as in the log-law region of a wall layer) is the mixing length related to a true eddy scale; otherwise it is just a quantity to be correlated empirically. Much the same applies to the eddy viscosity which, if transport terms are small, is the product of an eddy velocity scale and an eddy length scale (though it can be treated as one quantity).

An exact transport equation can be derived (from the Navier Stokes equations) for any length scale defined in terms of velocity fluctuations, such as the integral scale above. The equation can then be modelled in ways analogous to those outlined for the $\overline{u_i u_j}$ equations. If only *one* length scale is used to describe the turbulence it is implied that all relevant definitions are related to one another. Almost *none* of the terms in the exact length scale equation have ever been measured. These two facts imply an excessive number of degrees of freedom in the modelling of length scale transport equations.

Because the largest eddies in a turbulent shear layer always grow until they fill the shear layer we can, in simple cases, relate a real eddy length scale to the shear-layer thickness. This is more rigorous than relating the mixing length to the shear-layer thickness because in general the mixing length is not a true eddy length scale. At present the uncertainties of length scale modelling are such that a length scale related to the shear layer thickness, with empirical adjustments for special cases like stabilization by streamline curvature, *may* be more reliable than a length scale deduced from a transport equation.

Note that the length scale needed for modelling the Reynolds stress transport equation is a length scale of the larger eddies, which contain most of

Introduction

the contribution to $\overline{u_i u_j}$. When the turbulent energy dissipation rate ε is used in the definition of such a length scale it is really a substitute for the rate of turbulent energy transfer from the larger eddies to the smallest ones where the dissipation actually occurs. This energy transfer rate is of course equal to ε, but it depends on complicated triple correlations. The transport equation for energy transfer rate would be very complicated mathematically: the transport equation for ε itself is easier to write down but is ill-conditioned, depending on the small difference between two large terms.

Before passing on to the physics of turbulence, we briefly consider heat transfer, which is the quantity of primary interest in many engineering flows, although since it depends on the velocity field the latter must always be calculated first.

1.6 Heat transfer and compressible flow

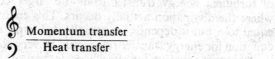

$$\dfrac{\text{Momentum transfer}}{\text{Heat transfer}}$$

- Momentum and heat transfer are caused by roughly similar mechanisms, but the differences are important
- Pollutant transfer ≈ heat transfer
- In adiabatic compressible flow

$$\frac{\rho'}{\rho} \approx -\frac{T'}{T} \approx (\gamma - 1)M^2 \frac{u}{U}$$

Heat transfer by turbulence is qualitatively similar to momentum transfer: however, the molecular diffusivity of heat (thermal conductivity) is different from the molecular diffusivity of momentum (viscosity), and pressure gradients appear in the momentum equation (panel 2.2) but not in the enthalpy equation shown on panel 2.3 (except in incompressible flow). It appears (Townsend, 1976; Launder, 1976) that Reynolds' quantitative analogy between heat-transfer parameters and momentum-transfer parameters is close only in the log-law part of a wall layer (panel 1.11): near free boundaries it is rather poor, though good enough for analogy factors to be useful subjects for data correlations.

A *musical* analogy exists: the relation between momentum transfer and heat transfer is akin to the relation between the treble and bass parts of a score—the bass contains roughly the same tune as the treble but there are important differences!

Passive scalar contaminants in small concentrations (small density variations) behave in the same way as enthalpy with small temperature differences; molecular diffusivities may differ but this seldom affects the turbulence much except in the "conductive sublayer" next to a solid surface.

Small density fluctuations in compressible flow, caused mainly by kinetic heating rather than by pressure fluctuations, appear to have little effect on the flow. (Mean density variations affect the mean-motion equations and may produce some changes in turbulence structure.) If pressure fluctuations are a small function of absolute pressure, the gas law gives the relation between density fluctuation ρ' and temperature fluctuation T' shown on the panel; the assumption of constant instantaneous total temperature (a strong version of Reynolds' analogy) leads to a relation between T' and the velocity fluctuation. In boundary layers $\sqrt{\overline{u^2}}/U \sim 0.05$ at low values of Mach number M and ρ'/ρ is small if $(\gamma - 1)M^2 \not> 1$ — say $M_e < 3$ to 5. In jets or mixing

Introduction

layers $\sqrt{\overline{u^2}}/U \sim 0{\cdot}2$, and compressibility effects on mixing rate start near $M = 1$.

Mass-weighted averages (see Cebeci and Smith, 1974, and Panel 15.4) remove density fluctuations from the equations—but not from the turbulence (see Bradshaw, 1977a).

1.7 Thin and fairly thin shear layers

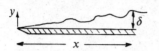
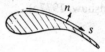

- Ratio of $\overline{\partial u^2}/\partial x$ to $\overline{\partial uv}/\partial y$ is $O(\delta/x)$, since $\overline{u^2} = O(\overline{uv})$ and $\partial/\partial x = O(\delta/x)\partial/\partial y$
- Therefore if δ/x or $d\delta/dx \ll 1$, x-component momentum equation is dominated by shear-stress gradient and the flow is a thin shear layer (TSL)
- Nearly all flows with significant Reynolds stress gradients are *fairly* thin shear layers, $\delta/x \not> 0.1$

A rigorous argument that gradients of Reynolds stress or viscous stress are significant only in shear layers is given by Cebeci and Bradshaw (1977): the only common exception is a shock wave. This implies that shear stress is the most important of the Reynolds stresses in most cases. In three-dimensional boundary layers of swept-wing type, the shear-stress vector is $(-\rho\overline{uv}, -\rho\overline{vw})$ taking y normal to the plane of the wing. In slender shear flows, as in wing-body junctions, shear-stress gradients still dominate the streamwise momentum equation but normal stress gradients affect the secondary flow (Chapter 11).

If $\partial/\partial x$ is not small with respect to $\partial/\partial y$ the rapid streamwise changes are probably due to strong pressure gradients so that—locally—Reynolds stresses do not matter.

Note that shear-layer arguments are most conveniently expressed in coordinates along and normal to the shear layer. In general these are semi-curvilinear (s, n) coordinates, but the differences between (x, y) and (s, n) coordinates are terms of order δ/R, where R is the radius of curvature. Usually R is of order x or larger and the δ/R terms are negligibly small in a thin shear layer.

If the shear layer is not thin enough for complete neglect of the terms of order δ/R or $d\delta/dx$ (as in the TSL equations) it will almost always be small enough for those terms to be approximated, perhaps crudely.

In the next few panels we consider the special features of shear layers before facing the problem of the mechanism of turbulence.

Introduction

1.8 Shear layers

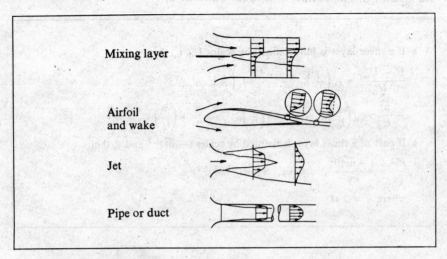

Except for the region near the trailing edge of the airfoil the flows shown on the panel are all thin layers with $\partial/\partial x \ll \partial/\partial y$, and either two-dimensional or axisymmetric. Turbulence models for flows like these are in a fairly satisfactory state, though some adjustment of empirical input for each flow may be advisable. These are the "building blocks" of most engineering turbulent flows; even rather complicated flows can be regarded qualitatively or quantitatively as assemblies of shear layers. For instance a wall jet can be treated as a mixing layer adjoining a boundary layer, while a separated flow "bubble" is bounded by one or more (highly distorted) mixing layers; the turbulence models required to describe these flows are likely to be more complicated and, as the complexity of the flow increases, less satisfactory.

1.9 Self-preservation and local equilibrium

- If a shear layer is fully defined by scales U_0, l_0, then
$$\frac{U - U_{\text{ref}}}{U_0} = f_1\left(\frac{y}{l_0}\right), \quad \frac{\partial U}{\partial y} = \frac{U_0}{l_0} f_1'\left(\frac{y}{l_0}\right),$$
$$-\frac{\overline{uv}}{U_0^2} = g\left(\frac{y}{l_0}\right), \quad \text{so } \frac{l}{l_0} = h\left(\frac{y}{l_0}\right), \quad \frac{v_\tau}{U_0 l_0} = \kappa\left(\frac{y}{l_0}\right)$$

- If part of a shear layer is defined by scales $(-\overline{uv})^{1/2}$ and y, then
$$\frac{\partial U}{\partial y} = \frac{(-\overline{uv})^{1/2}}{\kappa y}, \quad l = \kappa y, \quad v_\tau = (-\overline{uv})^{1/2} \kappa y$$

where $\kappa \approx 0.41$

Simple boundary conditions can lead to simple geometry and simple mathematical behaviour. *The eddies are just as complicated*—the apparent simplicity is explicable by dimensional analysis. If the flow can be specified by one length scale $l_0(x)$ then all lengths, whether mean flow scales, true eddy scales or hybrid scales like mixing lengths, are proportional to l_0. Examples include the far field of a circular jet ($l_0 \propto x$, $U_0 \propto 1/x$) and the *flat-plate boundary layer* ($l_0 = \delta$, $U_0 = U_\tau$).

Local equilibrium flows are a family with one member, for example the log-law part of a wall layer in mild pressure gradients. Here the length scale is y itself and the velocity scale is $U_\tau \equiv (\tau_w/\rho)^{1/2}$ or, more generally, $(-\overline{uv})^{1/2}$. The transport terms in the Reynolds stress transport equations are negligible, hence the name. It can be shown that the turbulent transport terms in a length-scale transport equation are *not* small. How could the length scale remain proportional to distance from the surface without transport of information from the surface? Thus the concept of local equilibrium should not be taken too far.

The viscous sublayer $U_\tau y/\nu < 30$ is not a local equilibrium region: turbulent transport of Reynolds stress towards the surface is significant because all Reynolds stresses fall to zero there.

Introduction

1.10 The viscous sublayer

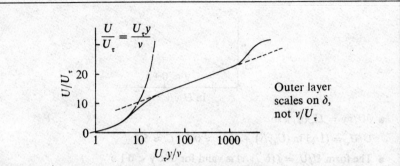

For $U_\tau y/\nu < 40$ (say), viscous stresses are significant. Total $\tau = \mu \partial U/\partial y - \rho \overline{uv} \approx \tau_w$. $l \approx \kappa y(1 - \exp[-U_\tau y/A^+ \nu])$ $U/U_\tau = f(U_\tau y/\nu)$

If U_τ and y are velocity and length scales of the larger, energy-containing eddies, $U_\tau y/\nu$ is an eddy Reynolds number and the effects of viscosity on the energy-containing eddies will be negligible if $U_\tau y/\nu \gg 1$. In the log-law region (viscous shear stress)/(total shear stress) = $1/(\kappa U_\tau y/\nu)$. For engineering purposes, mean viscous stress and other viscous effects are negligible for $U_\tau y/\nu > 30$–40.

For $U_\tau y/\nu < 40$, mixing length depends on $U_\tau y/\nu$. The variation shown (Van Driest, 1956), is best regarded as wholly empirical, giving a good fit to U. On a smooth solid surface with negligible $\partial \tau/\partial y$, $A^+ \approx 25$.

Turbulence affected by viscosity is bound to be more complicated than high Reynolds-number turbulence. Transport equation models can be extended into the low-Re range empirically but turbulence measurements are scarce. There is no good reason to suppose that a model adjusted to fit the viscous sublayer will also perform well in a transitional or reverse-transitional flow.

In the case of heat transfer, to which inner-layer similarity ideas can also be applied, there is a "conductive sublayer" in which molecular conductivity is important. Its thickness is roughly $40\nu/(U_\tau \text{Pr})$, where Pr is the molecular Prandtl number, of the order of 0·01 for liquid metals, 1 for gases, 10 for water and 100 or more for oils. Local equilibrium relations for heat transfer apply only outside *both* sublayers.

1.11 The logarithmic law

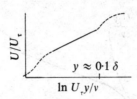

- $\partial U/\partial y = U_\tau/(\kappa y)$, so
  $U/U_\tau = (1/\kappa) \ln (U_\tau y/\nu) + C : \kappa \approx 0\cdot 41, C \approx 5\cdot 2$
- The form $U/U_\tau = f(U_\tau y/\nu)$ is valid for $0 < y < 0\cdot 1\, \delta$
- If $\tau = f(y)$ and $-\overline{uv} \approx \tau/\rho$
  $\partial U/\partial y = (-\overline{uv})^{1/2}/(\kappa y)$ to good approximation

For $30\text{--}40\, \nu/U_\tau < y < 0\cdot 1 - 0\cdot 2\delta$, $\tau_{\text{total}} \approx -\rho\overline{uv}$ and (panel 1.9)

$$\frac{\partial U}{\partial y} \approx \frac{(-\overline{uv})^{1/2}}{\kappa y}.$$

The relation is not exact if the flow is changing rapidly in the x direction, but seems to be quite accurate in the presence of large $\partial(-\overline{uv})/\partial y$, as in the case of transpiration with velocity V_w where $-\overline{uv} = U_\tau^2 + UV_w$. If $-\overline{uv}$ can be specified as a function of U, as here, or of y, the equation can be integrated from $U_\tau y/\nu = 30$ outwards and the velocity at $U_\tau y/\nu = 30$ inserted as an empirical "constant" of integration. The "constant" is actually a strong function of V_w/U_τ, the roughness Reynolds number $U_\tau k/\nu$, etc; dependence on $\partial\tau/\partial y$ via the parameter $\nu/(\rho U_\tau^3)(\partial\tau/\partial y)_w$ is small except very near separation. A fair body of data on the "constant" C, or on the equivalent constant in the Van Driest formula in panel 1.10, A^+, is available; see Cebeci and Smith (1974), Cebeci and Bradshaw (1977), Kader and Yaglom (1977) and Yaglom (1979).

At $y = 0\cdot 1 - 0\cdot 2\delta$, $U/U_e \sim 0\cdot 7$ in a boundary layer in small pressure gradient—so the log law does two-thirds of the calculations for us. In short regions of strong pressure gradients, the "mixing length" formula is reasonably accurate near the surface and the shear-stress gradient in the outer layer is small compared to the pressure gradient. Thus simple outer-layer turbulence models can succeed in many boundary layers or duct flows, even though they are incorrect in detail.

The next three panels discuss the behaviour of vorticity; this is the key to the inception and maintenance of turbulence.

Introduction

1.12 The mechanism of turbulence

$$\mathbf{v} = \tfrac{1}{4}\pi \Gamma \, d\mathbf{l} \wedge \mathbf{r}/r^3$$

where $\Gamma = \int \Omega \, dA$

Distortion of spanwise vortex line — "Horseshoe"

To proceed beyond the spurious simplicity of self-preserving flows or local equilibrium regions which can be correlated with a few empirical constants, we need to study the mechanism of turbulence.

Any fluid flow can be synthesized from a distribution of elementary vortex lines, on the analogy of magnetic lines of force. The top of the panel introduces the Biot–Savart law for induced velocity (which also applies to electromagnetic induction); the velocity in the sketch is directed into the paper, and depends on the vorticity Ω integrated over the cross-sectional area of the vortex. In turbulence the vorticity is largely concentrated into thin tubes or sheets, arranged nearly randomly and interacting via their induced velocity fields. The random velocity fluctuations cause two neighbouring points to move apart, on the average (Drunkard's walk theorem). If the two points lie on a vortex tube, the tube is stretched; its diameter decreases and, via conservation of angular momentum, kinetic energy of rotation passes to motions of smaller length scale. The small-scale vortex tube also extracts energy from the larger-scale motions that do the stretching, so the energy exchange effected by a given vortex tube operates over quite a large range of length scale. At the start of the process, energy is extracted from the mean flow via stretching of vortex lines by the mean strain rate. At the end, the energy of the very smallest vortex tubes or sheets, whose thickness is of order $10 \, (v^3/\varepsilon)^{1/4}$, is dissipated into heat by viscous stress fluctuations. The nearly random nature of this "cascade" of energy implies that any preferred orientation imposed by the mean strain rate does not reach the small-scale motion, which is therefore statistically isotropic to a first approximation. Therefore the shear stress is borne mainly by the large-scale motion.

An "eddy" is an assembly of strongly interacting vortex lines or sheets, interacting *comparatively* weakly with other eddies. The natural mode of distortion of a spanwise vortex line in a shear flow is a "horseshoe" vortex, with almost streamwise legs and an uplifted tip as shown in the lower part of the panel. This fits most observations of "bursts" and "large eddies." For a general introduction to turbulence mechanisms see Bradshaw (1975a) and for a more detailed treatment Bradshaw (1976). Intermediate between the two is Tennekes and Lumley (1972). Townsend (1976) and Monin and Yaglom (1971, 1975) give accounts which are extremely valuable to the expert but perhaps rather daunting to the beginner. The book by Rotta (1972a), which has unfortunately not been translated into English, achieves a helpful synthesis between the classical theory of homogeneous turbulence and the transport-equation approach to turbulence modelling.

1.13 The process of transition

1. Instability to infinitesimal (or small but finite) disturbances
2. Excitation of the instability by fluctuations in the external stream
3. Secondary instability (essentially nonlinear)
4. Nonperiodic components ("spectral broadening" or "jitter")
5. Turbulent spots
6. Merging of turbulent spots

Cebeci and Bradshaw (1977, Chapter 9) discuss instability and transition at some length. It can be explained by considering vortex dynamics (Bradshaw, 1976, Chapter 1). There is a gap between more or less exact solutions of the time-dependent equations for two-dimensional or simply three-dimensional disturbances, and empirical models for fully developed turbulence based on time-mean equations. The latter models could be extended upstream into the turbulent-spot region if the spot distribution at the start of the calculation were known, to make it well posed; even so, the empirical constants or functions in the calculation method would have to be made functions of Reynolds number. Extension into the pre-turbulent region would strain the models even further. However, a turbulence-model calculation started with a laminar-flow profile and a plausible velocity fluctuation level can qualitatively reproduce the growth of surface shear stress in transiton and may do so quantitatively, over a small range of cases at least, if adjusted to fit a set of test data: see, for example, McDonald and Fish (1973). The various mechanisms indicated on the panel are not separately represented by approaches of this type. An alternative approach, based on the solution of the Orr–Sommerfeld stability equation and also involving empiricism, is preferred here and is discussed in detail in Chapter 12.

1.14 Turbulent flow at low Reynolds number

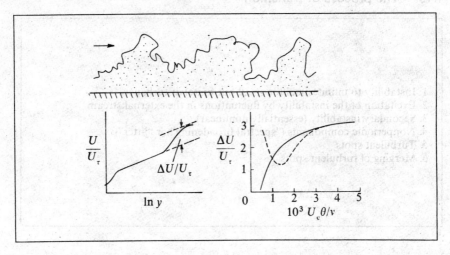

Vorticity passes from turbulent to non-turbulent fluid at a free-stream edge by means of viscous diffusion. The thickness of the viscosity-dominated "superlayer" is roughly the wavelength of the smallest turbulent eddies, but as the superlayer is spread over a highly-corrugated interface the volume occupied by the superlayer is quite large at low Reynolds number. Also, the decrease in the ratio of largest eddy size to smallest eddy size has some direct effect on large eddy structure. A simple measure of this in a boundary layer is the velocity defect or "wake" profile parameter, $\Pi \equiv \kappa \Delta U / 2 U_\tau$, which is constant in a zero pressure gradient flow at high Reynolds number. Most data correlate as shown by the full line. Large disturbances in the transition region, e.g. too big a trip wire, can produce results like the dotted line. Bushnell *et al* (1975) present data implying a *rise* in Π at low Re in boundary layers on supersonic wind tunnel walls, where transition occurs far upstream but Re_θ is kept low by acceleration in the nozzle. However low-speed experiments (Inman and Bradshaw, 1980) have failed to reproduce this rise. Low Reynolds number effects in jets appear only at low bulk Reynolds numbers, while there seem to be no low Reynolds number effects at all in the core region of duct flows (where there is no superlayer: Huffman and Bradshaw, 1972).

1.15 Real-life shear layers

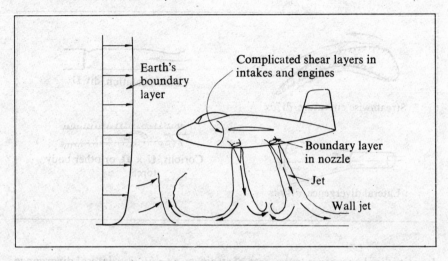

Here is a VTOL aircraft hovering in a headwind. The behaviour of the complex turbulent flows inside and outside the aircraft have been predicted to the satisfaction of the designer, but mainly by wind-tunnel or test-bed investigations rather than numerical calculations using turbulence models. However, calculation methods are starting to make a contribution even in highly complicated cases. Unfortunately, most of the complicated cases involve free shear layers, which are intrinsically harder to calculate than wall flows because of the absence of the universal similarity region of the wall layer. Self-preserving jets or wakes, unperturbed by obstacles or pressure gradients, can be predicted by simple data correlations. Non-self-preserving free shear layers, like those which bound separated-flow regions, are a different problem.

In the absence of a universal calculation method—and as an unavoidable part of the search for one—we need to study the response of shear layers to perturbations. A detailed review of perturbed boundary layers and internal flows is given by Fernholz and by Johnston (in Bradshaw 1976: Chapters 2 and 3). The next two panels illustrate the main types of perturbation and comment on their effect on the turbulence structure.

Examples of complex turbulent flows, involving recirculation, are discussed in Chapters 10 and 15 and show that numerical inaccuracies can, on occasion, be more important than those associated with the turbulence models.

1.16 Extra strain rates

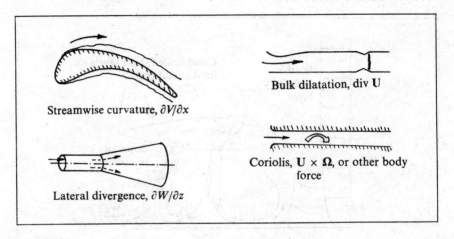

Longitudinal curvature (extra rate of strain $\equiv e = \partial V/\partial x$), lateral divergence ($e = \partial W/\partial z$), buoyancy, system rotation ("e" = Ω), bulk dilatation ($e = -\operatorname{div} \mathbf{U}$), and possibly other examples of rate of strain or body-force components extra to the basic shear $\partial U/\partial y$, can have a large effect on turbulence. Of course the Navier–Stokes equations still apply, but most current calculation methods are such crude simplifications of the Navier–Stokes equations that they will not reproduce extra strain effects without extra empirical input. Typically the explicit extra terms that appear in the Reynolds stress transport equations are of order $e/(\partial U/\partial y)$ times the original generation terms containing the mean strain rate: the fractional change in eddy viscosity, however, is typically $10 e/(\partial U/\partial y)$.

Responses to sudden changes in extra rate of strain, such as the passage of a shear layer through a short region of large curvature, are complicated, and the return to plane flow conditions can be non-monotonic. In general, flows with extra strain rates are severe tests of calculation methods and these and shear-layer interactions, as two main classes of complex flow, are discussed by Bradshaw (1975b): longitudinal curvature has been considered at length by Bradshaw (1973).

Introduction

1.17 Interactions

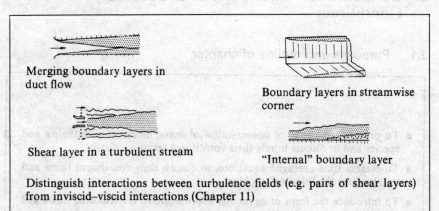

Merging boundary layers in duct flow

Boundary layers in streamwise corner

Shear layer in a turbulent stream

"Internal" boundary layer

Distinguish interactions between turbulence fields (e.g. pairs of shear layers) from inviscid–viscid interactions (Chapter 11)

In asymmetrical flows $\partial U/\partial y$ and $-\overline{uv}$ do not go through zero at the same point. This implies that the eddy viscosity goes to infinity and then becomes negative while the mixing length goes to infinity and then becomes imaginary. Qualitatively one can see that an asymmetrical flow consists of two slightly different, interacting simple shear layers. Symmetry disguises the more spectacular consequences of the interaction since the eddy viscosity remains finite (although l goes to ∞ as $(y - y_{\text{CL}})^{-1/2}$). A Reynolds stress transport model should remain well behaved even in an asymmetrical flow; however $(-\overline{uv})^{1/2}$ is no longer a suitable velocity scale since it goes to zero in the middle of the turbulent region. Two advantages of treating the interaction as two overlapping turbulence fields are (i) that the shear stress carried by one layer can be used as an eddy velocity scale for that layer only, and (ii) that the eddy length scale can be related to the thickness of that layer only, avoiding the need for a length-scale transport equation with its attendant uncertainty and increased computing cost. Bradshaw *et al.* (1973) calculated fully developed duct flow using only boundary-layer data: in more highly turbulent flows the turbulence structure of each layer, and thus the empirical data in the calculation method, is changed by the interaction. The wake of an airfoil (Andreopoulos and Bradshaw, 1980) is a special case because the two boundary layers interact on their low-velocity sides where the eddies are very small.

2 Conservation Equations and Boundary Conditions

2.1 Purposes and outline of chapter

- To present equations of conservation of mass, momentum, enthalpy and species and to discuss briefly their constituent terms
- To present time-averaged equations, to discuss their constituent terms and to introduce the "closure" problem
- To introduce the form of equations appropriate to recirculating flows and thin-shear layers and to discuss boundary condition requirements

---

The equations of conservation of mass, momentum, enthalpy and species are considered in the first section of this chapter. Their basis is described briefly and important functions are to introduce terminology and to provide a background of "exact" equations against which modelled equations can be judged. Fluid properties, other than those mentioned above, can be conserved and represented by related equations. The partial differential equations are discussed term by term and, although the equations are presented in tensor notation and in rectangular Cartesian coordinates, alternative forms are discussed. This discussion continues that presented in relation to panels 1.2 to 1.4.

The reasons for time- or ensemble-averaging the equations are presented and the time-averaged equations discussed in a parallel manner to that for the time-dependent equations. Particular attention is paid to the two-dimensional, time-averaged equations in the form corresponding to recirculating and boundary layer flows and the boundary condition requirements are discussed.

2.2 Navier–Stokes equations

$$\frac{\partial \hat{\rho}}{\partial t} + \frac{\partial (\hat{\rho} \hat{U}_i)}{\partial x_i} = 0$$

$$\frac{\partial \hat{U}_i}{\partial t} + \hat{U}_j \frac{\partial \hat{U}_i}{\partial x_j} = -\frac{1}{\hat{\rho}} \frac{\partial \hat{P}}{\partial x_i} + \frac{1}{\hat{\rho}} \frac{\partial}{\partial x_j} \left(\mu \frac{\partial \hat{U}_i}{\partial x_j} \right) + \frac{1}{3\hat{\rho}} \frac{\partial}{\partial x_i} \left(\mu \frac{\partial \hat{U}_l}{\partial x_l} \right)$$

$\hat{U}$, $\hat{P}$ and $\hat{\rho}$ are functions of time

The equations of conservation of mass and momentum are shown in the panel and may be obtained from a control-volume analysis and, in the case of the momentum equation, from Newton's viscous-stress law, $\tau = \mu(\partial U/\partial y)$ and its generalizations. The continuity equation, in expanded form, has one term corresponding to the rate of change of bulk density and three terms corresponding to the spatial gradients of mass flow.

In the momentum equation the rate of change of velocity is balanced by convection, pressure gradient and diffusion. The convective term may be expanded as

$$\left(\hat{U}_1 \frac{\partial \hat{U}_1}{\partial x_1} + \hat{U}_2 \frac{\partial \hat{U}_1}{\partial x_2} + \hat{U}_3 \frac{\partial \hat{U}_1}{\partial x_3} \right)$$

with similar terms in the $\hat{U}_2$ and $\hat{U}_3$ equations. The diffusion term, for constant viscosity and density, has the form

$$\nu \left(\frac{\partial^2 \hat{U}_1}{\partial x_1^2} + \frac{\partial^2 \hat{U}_1}{\partial x_2^2} + \frac{\partial^2 \hat{U}_1}{\partial x_3^2} \right)$$

with similar terms in the $\hat{U}_2$ and $\hat{U}_3$ equations.

With simple boundary conditions, reduced forms of these equations can be solved exactly and by similarity methods. In general, they can be solved numerically but, as will be discussed, only for low values of Reynolds number.

2.3 Enthalpy

- $$\frac{\partial \hat{H}}{\partial t} + \hat{U}_j \frac{\partial \hat{H}}{\partial x_j} = \frac{\hat{S}_H}{\hat{\rho}} + \frac{1}{\hat{\rho}} \frac{\partial}{\partial x_j} \left(k \frac{\partial \hat{T}}{\partial x_j} \right)$$

- $\hat{U}_j$, $\hat{H}$, $\hat{T}$ and $\hat{\rho}$ are functions of time

  $\hat{S}_H$ may represent, for example, shear work or radiation

  $$\hat{H} \equiv C_p \hat{T} + \frac{\hat{U}_i^2}{2} + \hat{M}_{fu} \hat{H}_{fu}$$

- for an ideal fluid of uniform thermal conductivity

  $$\frac{\partial \hat{T}}{\partial t} + \hat{U}_j \frac{\partial \hat{T}}{\partial x_j} = \frac{\hat{S}_H}{\hat{\rho} C_p} + \frac{k}{\hat{\rho} C_p} \frac{\partial^2 \hat{T}}{\partial x_j^2}$$

Conservation of total enthalpy, as defined in the panel, together with Fourier's law of heat conduction, i.e. $\dot{q}_x'' = -k(\partial T/\partial x)$, leads to the first equation which balances the rate of change of total enthalpy with convection, sources and diffusion. The derivation of this equation or the reduced form of the energy equations shown in the panel is achieved by a control volume analysis similar to that which results in the continuity and momentum equations. As pointed out in relation to panel 1.6, the form of the equation is similar to that of the momentum equation although $\hat{H}$, in contrast to $\hat{U}_j$, is a scalar quantity. In particular, if the source terms are absent, the momentum and enthalpy equations are identical in form and the solution method for one will also be satisfactory for the other. The two equations will be coupled, if the temperature differences are large enough to cause significant variations in density; otherwise the uncoupled equations can be solved sequentially. Only in special cases will the source term, $\hat{S}_H$, appear. Shear work, for example, is a high Mach number phenomenon and radiation, which will be discussed briefly in Chapter 15, is particularly relevant to large-scale combusting flows. Except in reacting flows, $\hat{M}_{fu}\hat{H}_{fu}$, the product of the mass fraction of fuel and its enthalpy of combustion for unit mass, is ignored in the definition of $\hat{H}$.

2.4 Species equation

For a two-component mixture

$$\frac{\partial \hat{\rho}_a}{\partial t} + \frac{\partial}{\partial x}(\hat{\rho}_a \hat{U}) + \frac{\partial}{\partial y}(\hat{\rho}_a \hat{V}) + \frac{\partial}{\partial z}(\hat{\rho}_a \hat{W}) = \hat{M}_a +$$

$$\frac{\partial}{\partial x}\left(\rho \mathscr{D}_{ab} \frac{\partial \hat{m}_a}{\partial x}\right) + \frac{\partial}{\partial y}\left(\rho \mathscr{D}_{ab} \frac{\partial \hat{m}_a}{\partial y}\right) + \frac{\partial}{\partial t}\left(\rho \mathscr{D}_{ab} \frac{\partial \hat{m}_a}{\partial z}\right)$$

The equation of continuity for species a is a formulation, in mathematical terms, of the statement that the rate of accumulation of a in a fixed region in space is equal to the excess of flow rate into the region over the flow rate out, plus the rate of production of a in the region. Control volume analysis leads to

$$\frac{\partial \hat{\rho}_a}{\partial t} + \frac{\partial \hat{G}_{a,x}}{\partial x} + \frac{\partial \hat{G}_{a,y}}{\partial y} + \frac{\partial \hat{G}_{a,x}}{\partial z} = \hat{M}_a$$

where $\hat{\rho}_a$ is the mass density of species a, $\hat{G}_{a,x}$ is the mass of a crossing exit area $(\hat{\rho}_a U_i)$ per unit time and $\hat{M}_a$ is the mass rate of creation of a per unit volume per unit time. Note that the continuity equation for the mixture can be obtained from:

$$\sum_a \hat{\rho}_a = \hat{\rho}; \quad \sum_a \hat{G}_{a,x} = \hat{\rho}U; \quad \sum_a \hat{M}_a = 0.$$

Fick's law of diffusion,

$$\hat{\rho}_a(U_a - U) = -\mathscr{D}_{ab}\frac{\partial \hat{m}_a}{\partial x},$$

completes the necessary information and leads to the equation shown on the panel.

2.5 Averaged momentum equation

With $\hat{U} = U + u$, $\hat{P} = P + p$ and neglecting correlations with density fluctuations, the assumption that

$$U = \lim \frac{1}{t_1 - t_2} \int_{t_2}^{t_1} \hat{U}(x_1, x_2, x_3, t) \, dt$$

$$(t_1 - t_2) \to \infty$$

and the equations of panel 2.2 lead to

$$\frac{\partial}{\partial x_i} \overline{\rho \hat{U}_i} = 0$$

$$U_j \frac{\partial}{\partial x_j} U_i = -\frac{1}{\rho} \frac{\partial P}{\partial x_i} + \frac{1}{\rho} \frac{\partial}{\partial x_j}\left(\mu \frac{\partial U}{\partial x_j}\right) - \frac{\partial}{\partial x_j} \overline{u_i u_j}$$

If the time-dependent equations of the previous panels could be solved for general boundary conditions, the problems of fluid mechanics, heat and mass transfer would be rapidly solved. This cannot be achieved by analytic means and discretized methods offer the only possibility of achieving this aim. Unfortunately, turbulent flows contain fluctuation energy at frequencies from zero to 10 kHz or more, with a corresponding wide range of wavelengths, which implies the need for a very fine finite-difference mesh in the three orthogonal directions (turbulence is three-dimensional) and in time. As a consequence of this and available computers, numerical solutions have been restricted to low Reynolds numbers and to relatively simple boundary conditions.

One possible, and popular, way of reducing the magnitude of the problem is to time- or ensemble-average the equations and reduce the number of independent variables. The last equation results, with the added simplifying assumption of uniform density, and contains the Reynolds stress term. Thus, the equation set has more unknowns than equations. Apart from the additional Reynolds stress term, and the absence of the time-dependent term, the equation is similar in form to the original equation.

The continuity equation has been written with an overall average to remind us that, if ρ fluctuates, the term $\partial/\partial x_i \overline{\rho' u_i}$ cannot be neglected unless the correlation is known to be small. In high Mach number and combusting flows, this will not be the case.

2.6 Averaged general scalar equation

> Corresponding average equations may be obtained for enthalpy and species: they have the form
> $$U_j \frac{\partial}{\partial x_j}\Phi = S_\Phi + \frac{\partial}{\partial x_j}\left(\Gamma_\Phi \frac{\partial \Phi}{\partial x_j}\right) - \frac{\partial}{\partial x_j}\overline{u_j \phi}$$
> where Φ and ϕ are the mean and fluctuating parts of the scalar being considered, and Γ_Φ is its diffusivity.

The Reynolds stress term has its counterpart in the averaged equations for enthalpy and species; thus the term $\overline{u_j \phi}$ represents the local heat or mass flux respectively. Once again, the equation is insoluble in that Φ and $\overline{u_j \phi}$ are both unknown. Experiments in turbulent flows have shown that, in both the equations for the vector and scalar quantities, the magnitude of the turbulent diffusion term is substantially greater than the laminar-diffusion term except in the immediate vicinity of solid surfaces. Thus, the unknown correlations cannot be neglected.

It should be realized that the equation in the panel is valid for dependent variables other than enthalpy and species. Scalar quantities such as, for example, particle or droplet number density can also be represented in this form.

The problem of the unknown correlations may be solved, in principle, by representing them as dependent variables in equations of the form in the panel: this is discussed in the following panel. It may also be solved by specifying the correlations as functions of known quantities as, for example, by Prandtl's mixing length for $\overline{u_1 u_2}$, and the turbulent Prandtl number approach described in Chapter 3.

2.7 Reynolds stress equation

For a fluid of uniform viscosity and density, unaffected by gravitational, magnetic or other external forces, the transport of Reynolds stress is governed by the equation:

$$\frac{D\overline{u_i u_j}}{Dt} = -\left(\overline{u_i u_j}\frac{\partial U_j}{\partial x_k} + \overline{u_j u_k}\frac{\partial U_i}{\partial x_k}\right) - 2\nu \overline{\frac{\partial u_i}{\partial x_k}\frac{\partial u_j}{\partial x_k}} + \overline{\frac{p}{\rho}\left(\frac{\partial u_i}{\partial x_j} + \frac{\partial u_j}{\partial x_i}\right)}$$
$$\text{(I)} \qquad\qquad\qquad\qquad \text{(II)} \qquad\qquad \text{(III)}$$

$$-\frac{\partial}{\partial x_k}\left(\overline{u_i u_j u_k} + \delta_{ik}\frac{\overline{u_i p}}{\rho} + \delta_{ik}\frac{\overline{u_j p}}{\rho} - \nu \frac{\partial \overline{u_i u_j}}{\partial x_k}\right)$$
$$\text{(IV)}$$

The equation representing the transport of Reynolds stress may readily be derived as described in connection with panel 1.3 and is shown above. The closure difficulty of panel 1.4 is confirmed by the presence of the triple correlation. The pressure correlations and dissipation terms are additional unknowns and the equation is insoluble without further information.

The left-hand side of the equation represents convection and does not introduce further unknowns; like the generation term (I), it does not require to be "modelled" and can, in principle, be measured. The dissipation correlation (II) has neither of these desirable features but may be represented (assuming local isotropy of the dissipating eddies) in the form

$$\varepsilon\, \overline{u_i u_j}/k$$

(see Donaldson, 1969; Daly and Harlow, 1970), where $\varepsilon \equiv \nu(\partial u_i/\partial x_j)^2$ and k is the turbulence kinetic energy. The pressure–strain correlation (III) and the diffusion transport terms (IV) imply that, in addition to information on k and ε, assumptions are needed to relate these terms to known quantities. The exact definition of the dissipation rate is given in panel 1.5. It is equal to ε, as defined here, if dissipating eddies are isotropic, which is true except in or near the viscous sublayer.

2.8 Turbulent-energy equation

The equations representing transport of normal stresses are

$$U_j \frac{\partial}{\partial x_j} \overline{u_{(i)}^2} + 2\overline{u_{(i)}u_j} \frac{\partial U_{(i)}}{\partial x_j} + \frac{\partial}{\partial x_j}[\overline{u_{(i)}^2 u_j} + \overline{2u_{(i)}p\,\delta_{(i)j}}]$$

$$- \frac{2}{\rho}\overline{p \frac{\partial u_{(i)}}{\partial x_{(i)}}} - 2\nu \frac{\overline{\partial^2 u_{(i)}^2}}{\partial x_k^2} + 2\varepsilon = 0$$

$$k = \overline{(u_1^2 + u_2^2 + u_3^2)}/2 = \overline{u_i^2}/2$$

The individual normal-stress equations may be derived from the Reynolds stress equations by taking $i = j$ *without* summing, and the equation for turbulent kinetic energy is obtained by summing the normal stresses over the indices in parentheses.

Once again, the convection and generation terms contain dependent variables of equations already discussed. The triple correlation which corresponds to the transport of turbulent kinetic energy by the turbulence—a form of diffusion—is unknown. The pressure diffusion term is also unknown; the term represents the gain of turbulent energy of the flow by pressure gradients. It is useful to note that the pressure strain term of the Reynolds stress equation has disappeared. The last term may be divided into two parts which represent diffusion of turbulent kinetic energy by viscous forces and dissipation; the former is small except near walls and the latter represents an unknown for which a conservation equation may be derived.

The turbulent energy is distributed over a wide range of eddy sizes and energy spectrum equations may also be derived; such equations introduce the additional complexity of spectral and spatial transport with corresponding additional unknowns. At present, one equation for typical eddy length scale is regarded as sufficient. The next two panels show how such an equation can be obtained.

2.9 Dissipation equation

$$\frac{\partial^2 U_i}{\partial x_l \partial x_j}\overline{u_j \frac{\partial u_i}{\partial x_l}} + \frac{\partial U_i}{\partial x_j}\overline{\frac{\partial u_i}{\partial x_l}\frac{\partial u_j}{\partial x_l}} + \frac{\partial U_j}{\partial x_l}\overline{\frac{\partial u_i}{\partial x_l}\frac{\partial u_i}{\partial x_j}} + U_j\frac{\partial}{\partial x_j}\overline{\left\{\frac{1}{2}\left(\frac{\partial u_i}{\partial x_l}\right)^2\right\}}$$
$$\text{(a)} \qquad \text{(b)} \qquad \text{(c)} \qquad \text{(d)}$$

$$+ \overline{\frac{\partial u_i}{\partial x_l}\frac{\partial u_i}{\partial x_j}\frac{\partial u_j}{\partial x_j}} + \overline{u_j \frac{\partial}{\partial x_j}\left\{\frac{1}{2}\left(\frac{\partial u_i}{\partial x_l}\right)^2\right\}} = -\frac{1}{\rho}\overline{\frac{\partial u_i}{\partial x_l}\frac{\partial^2 p}{\partial x_l \partial x_i}} + \nu \overline{\frac{\partial u_i}{\partial x_l}\left(\frac{\partial^2}{\partial x_j^2}\frac{\partial u_i}{\partial x_l}\right)}$$
$$\text{(e)} \qquad \qquad \text{(f)} \qquad \qquad \text{(g)} \qquad \qquad \text{(h)}$$

As suggested on the previous page and made clear by its presence in the Reynolds stress equation, the dissipation term is unknown and may be expressed as the dependent variable of a differential conservation equation. The equation in the panel has, as dependent variable

$$\varepsilon \equiv \nu \overline{(\partial u_i/\partial x_j)^2}$$

which is the main part of the full dissipation term: see panels 1.5 and 2.7.

The above equation is obtained from the Navier–Stokes equation written in terms of $(U_i + u_i)$ and the corresponding continuity equation. The time average forms are

$$U_j \frac{\partial U_i}{\partial x_j} + \overline{\frac{u_j \partial u_i}{\partial x_j}} = -\frac{1}{\rho}\frac{\partial P}{\partial x_i} + \nu \frac{\partial^2 U_i}{\partial x_j^2}$$

and

$$\frac{\partial U_i}{\partial x_i} = 0.$$

With the Navier–Stokes equations, these may be rewritten in the form

$$\frac{\partial u_i}{\partial t} + u_j \frac{\partial (U_i + u_i)}{\partial x_j} + \frac{U_j \partial u_i}{\partial x_j} = -\frac{1}{\rho}\frac{\partial p}{\partial x_i} + \frac{\partial^2 u_i}{\partial x_j^2} + \frac{\partial}{\partial x_j}\overline{u_i u_j}.$$

Differentiation of this equation with respect to x_l and multiplication by $\partial u_i/\partial x_l$ leads to an equation which is then ensemble averaged to yield the equation of the panel.

Convection and dissipation are in recognisable form but otherwise the equation is obscure and brings in many unknown correlations. It may be contrasted with the form proposed by Daly and Harlow (1970),

$$U_i \frac{\partial \varepsilon}{\partial x_i} = \nu \nabla^2 \varepsilon + 2 \frac{\partial}{\partial x_i}\left(\frac{k}{\varepsilon}\overline{u_i u_j}\frac{\partial \varepsilon}{\partial x_j}\right) + 2 \frac{\varepsilon}{k}\frac{\partial}{\partial x_i}\left(\frac{k}{\varepsilon}\overline{u_i u_j}\frac{\partial k}{\partial x_j}\right)$$

Convection   Molecular diffusion   Turbulent diffusion

$$- C_1 \frac{\varepsilon}{k}\overline{u_i u_j}\frac{\partial U_i}{\partial x_j} - C_2 \frac{\varepsilon^2}{k}$$

Production   Dissipation

To arrive at this equation, term (a) was neglected, terms (b) and (c) combined to represent production and (e), (f) and (g) combined to represent diffusion.

2.10 Length scales

- To describe a turbulent flow we need length scales as well as velocity scales: a length-scale equation can be obtained from the equation of the previous panel with

$$L_\varepsilon \equiv (\overline{u^2}/2)^{3/2}/\varepsilon$$

- Macro- and micro-scales can be obtained, in principle, from solutions of a transport equation for L_x

---

Turbulent flows are partly characterised by the scale of turbulence and corresponding equations are of relevance to numerical formulations. As indicated in panel 1.5, a dissipation length scale may be obtained from the dissipation equation of the previous panel if local isotropy of the dissipation eddies is assumed. The resulting equation, in terms of $k^{3/2}/L_\varepsilon$, has direct physical significance — alternative forms, $k^m L^n$, can also be derived (panel 1.4) but with less direct physical implications.

Macro- and micro-scales can also be obtained from the double-velocity correlation function, with spatial separation in the x-direction, say.

$$R(x, r, \tau) = \overline{u_i(x, t) u_j(x + r, t + \tau)}$$

with
$$L_x = \int_{-\infty}^{+\infty} R(x, r, 0)\, dr, \quad \text{and} \quad T_x = \int_{-\infty}^{\infty} R(x, 0, t)\, dt.$$

A micro-scale may be obtained from the second gradient of $R(x, r, 0)$ with respect to r at $r = 0$.

The equation for L_x is obtained by writing the Navier–Stokes equations for $U_i + u_i$ at point A, multiplied by u_j at point B: adding the equation for $U_j + u_j$ at point B multiplied by u_i at A: this, after manipulation, leads to an equation in terms of $R_{ij}(r) = \overline{u_i^A u_j^B}$. As might be expected from previous discussions of the "closure problem," it includes terms of the form $\overline{u_i^A u_k^A u_j^B}$.

2.11 Scalar flux equations

$$\frac{D\overline{u_i\phi}}{Dt} = -\left(\overline{u_i u_k}\frac{\partial \Phi}{\partial x_k} + \overline{\phi u_k}\frac{\partial U_i}{\partial x_k}\right) - \frac{\overline{\alpha\phi^2 g_i}}{\Phi} - (\lambda + \nu)\overline{\frac{\partial \phi}{\partial x_k}\frac{\partial u_i}{\partial x_k}} + \overline{\frac{p}{\rho}\frac{\partial \phi}{\partial x_i}}$$
$$\phantom{\frac{D\overline{u_i\phi}}{Dt} =}\text{Production}\qquad\qquad\text{Buoyancy}\quad\text{Dissipation}\quad\text{Pressure}$$
$$\phantom{\frac{D\overline{u_i\phi}}{Dt} =}\text{(I)}\qquad\qquad\qquad\text{(II)}\qquad\quad\text{(III)}\qquad\quad\text{(IV)}$$

$$-\frac{\partial}{\partial x_k}\left(\overline{u_k u_i \phi} + \frac{\overline{p\phi}}{\rho}\delta_{ik}\right)$$
$$\text{Diffusion (V)}$$

The derivation of an equation for the Reynolds stress correlation found in the averaged momentum equations may be paralleled by an equation for the scalar flux term, $\overline{u_j\phi}$ of panel 2.6. The result is shown in the panel; viscous diffusion has been omitted from the equation but the buoyant-force contribution has been included. The convection and production terms of the equation are obvious as in the buoyancy term. If the small-scale dissipative eddies are presumed isotropic, term (III) will disappear but the pressure scalar term (IV) is significant and, as in the Reynolds stress equation, introduces an unknown term. The diffusion term (V) also introduces unknown quantities. It should be realized that, if the buoyancy term of the above equation is significant, the Reynolds stress equation (and the mean flow equations) may require corresponding sources.

The equations of panels 2.7, 2.8, 2.9 and 2.11 cannot be solved in the forms presented and modelling assumptions are required before they can be useful to numerical solution methods. Appropriate assumptions and the consequent equations are discussed in Chapters 3 and 4. As will be shown, each of these equations with modelling assumptions can be written in the form of the averaged general scalar equation of panel 2.6 so that one solution method can be used for equations with dependent variables including $\overline{u_i u_j}$, k, ε and $\overline{u_i\phi}$.

2.12 Two-dimensional elliptic equations

Averaged equations, for a steady two-dimensional flow, may be written in the form

$$\rho \frac{\partial}{\partial x}(U\Phi) + \frac{\rho}{r}\frac{\partial}{\partial y}(rV\Phi) - \frac{\partial}{\partial x}\left(\Gamma_\Phi \frac{\partial \Phi}{\partial x}\right) - \frac{1}{r}\frac{\partial}{\partial y}\left(r\Gamma_\Phi \frac{\partial \Phi}{\partial y}\right) = S_\Phi$$

with $\Phi = U, V, W, \overline{u_i u_j}, \varepsilon, H, M_a$, etc.
Γ_Φ = diffusion coefficients, e.g. μ_{eff}
S_Φ = source, e.g. $-\dfrac{\partial P}{\partial x}$
r = 1 for Cartesian coordinates and $r = y$ for cylindrical coordinates

---

In order to allow the solution of the equations presented on previous panels, it is desirable that they be represented in a single form and, therefore, solved by a single numerical scheme. The equation in the panel represents a two-dimensional, elliptic form (see Chapter 5) in which the previous equations—in modelled form—will be represented. Thus, for example, the averaged Navier–Stokes equations for a two-dimensional flow have convective and diffusion terms similar to those in the panel; the resulting source term can include turbulent diffusion and additional terms. In a modelled form of this equation, the diffusion terms can represent effective diffusivity (see Chapter 3).

Equations of this elliptic form require the specification of boundary conditions on all sides of the solution domain. Thus, for example, the U-momentum equation requires that the U-velocity or its gradient be specified on all boundaries. Both U- and V-momentum equations (with continuity) are needed in regions where the pressure gradient and diffusion terms in orthogonal directions have similar magnitudes. In general, recirculating flows ("separated") are of this kind and are characterised by "upstream influence," i.e. downstream regions of flow influencing upstream regions of flow; hence the need to take account of downstream boundary conditions. In many cases a separated flow will relax into a boundary-layer flow and the downstream boundary condition can be represented as $\partial \Phi/\partial x = 0$ at a location sufficiently far downstream such that the calculations in the region of recirculation are uninfluenced.

Conservation Equations and Boundary Conditions

2.13 Two-dimensional, parabolic equations

- Averaged equations, for a two-dimensional boundary-layer (or thin shear-layer) flow may be written in the form

$$\rho \frac{\partial}{\partial x}(U\Phi) + \frac{\rho}{r}\frac{\partial}{\partial y}(rV\Phi) - \frac{1}{r}\frac{\partial}{\partial y}\left(r\Gamma_\Phi \frac{\partial \Phi}{\partial y}\right) = S_\Phi$$

- This equation is parabolic and boundary conditions are required on three sides of the solution domain

For a thin shear layer, in which there is a preferred direction of flow and with diffusion much larger in the cross-stream direction, order-of-magnitude arguments allow the equation of panel 2.12 to be reduced to the above. The familiar momentum equation, with μ_{eff} representing viscosity and turbulence effect,

$$\rho U \frac{\partial U}{\partial x} + \rho V \frac{\partial U}{\partial y} - \frac{\partial}{\partial y}\left(\mu_{\text{eff}} \frac{\partial U}{\partial y}\right) = -\frac{dP}{dx}$$

is of this form with equivalent equations representing, for example, enthalpy, species concentration, Reynolds stress etc.

The parabolic nature of the equations (see Chapter 5) implies that the flow is uninfluenced by downstream events and that boundary conditions are required on only three sides of the solution domain. Solution procedures for equations of this form are generally of the "marching" type, in contrast to the "iteration" methods necessary for elliptic equations.

In unsteady flows with recirculation, and in corner flows where there is a preferred flow direction *and* recirculation in the plane orthogonal to this direction, the numerical scheme combines iteration in the "elliptic" plane with marching in time or in the direction of the flow.

2.14 Concluding remarks

The properties of turbulent flows would be known if exact equations, such as those of panels 2.2 and 2.3 could be solved with the appropriate boundary conditions. Unfortunately, the large range of time and spatial scales present in turbulent flows, together with the limited storage of digital computers, precludes this possibility. The equations may be simplified by time (or ensemble) averaging with the immediate consequences of the "closure problem" associated with the additional correlation term and the knowledge that any solution will be approximate and require some confirmation from measurement. Equations can be derived with the unknown correlations as dependent variables and always involve additional unknowns including a higher-order correlation term.

The equations considered were all presented without the modelling assumptions required to reduce the number of unknowns to the number of equations, which are the subject of Chapters 3 and 4. It is useful to note, however, that the various equations can be written in similar form and it is therefore to be expected that, for a given flow problem, the equations for conservation of momentum, enthalpy, species etc. can be solved by the same numerical technique.

3 Turbulence Models Based on Eddy Viscosity Hypotheses

3.1 Purposes and outline of chapter

- To present equations corresponding to eddy-viscosity-type closures and to relate them to the exact equations of Chapter 2
- To assess, briefly, by consideration of the assumptions, the likely validity of the eddy viscosity methods
- Particular attention is devoted to two-equation models, to low Reynolds number formulations and to algebraic stress models
- The turbulent "Prandtl" number approach to the modelling of scalar equations is considered in a similar manner

The equations presented in Chapter 2 cannot be solved in their exact form, and this and the following chapters present and discuss modelled forms and corresponding closures. This chapter is concerned with eddy viscosity methods and the following one with Reynolds stress closures. The sequence has been chosen to allow the introduction, first, of simple and comparatively well-known procedures. The reverse sequence would be equally logical in that the equations and models of this chapter can be derived from the Reynolds stress approach.

Attention is also devoted to the turbulent "Prandtl" number method of representing scalar fluxes. In general, the approaches discussed here are of current engineering value whereas those of the following chapter are of more fundamental interest.

3.2 Eddy viscosity concept, 1

- In near-equilibrium shear flows, the kinetic energy equation is dominated by the source terms, i.e. production and dissipation

  Thus $\quad 1 \simeq P/\varepsilon = \overline{u_i u_j}\dfrac{\partial U_i}{\partial x_j}/\varepsilon$

  and we can define a realistic eddy length scale, as in panels 1·5 and 2·10, by $L_\varepsilon \equiv k^{3/2}/\varepsilon$ where $k \equiv \overline{u_i^2}/2$
  Therefore, $\overline{u_i u_j} = f(\partial U_i/\partial x_j, k, l)$

- In a simple shear layer, $P = -\overline{uv}\partial U/\partial y$ and the mixing length l (panel 1·5) is $(-\overline{uv}/k)^{3/2} L_\varepsilon$

- In addition, we have Boussinesq's concept of eddy viscosity,

  $$-\overline{u_i u_j} = v_t\left(\dfrac{\partial U_i}{\partial x_j} + \dfrac{\partial U_j}{\partial x_i}\right) - \tfrac{2}{3}\delta_{ij}k$$

The eddy viscosity concept stems from the convenience associated with maintaining an approach, for turbulent flows, which is similar to that for laminar flows. This convenience of concept is coupled with the possible mathematical convenience of retaining the same form of differential equations for laminar and turbulent flow and allowing the use of the same solution procedure.

The panel indicates Boussinesq's eddy viscosity formulation, together with an argument which leads to the conclusion that the shear stress is proportional to mean velocity gradient, turbulence energy and a length scale characteristic of energy-bearing eddies. Two assumptions are required to reach this useful conclusion. The first is that the flow is not far from equilibrium so that convection and diffusion are small compared to production and dissipation and the second is that the turbulence Reynolds number is high. These assumptions may be regarded as an indication of the limitations of the eddy viscosity approach.

In panels 4.4 and 4.5, the near-equilibrium assumption is shown to be approximately correct for a mixing layer and, for example, the wall boundary-layer results of Klebanoff (1955) also suggest a similar trend: in both cases, the flow properties change relatively slowly from one downstream location to another and the flow can be said to be nearly in equilibrium.

3.3 Eddy viscosity concept, 2

Levels of closure:

- zero equation; $v_t = c_\mu l_1^2 \left| \left(\dfrac{\partial U_i}{\partial x_j} + \dfrac{\partial U_j}{\partial x_i} \right) \right|$: l from algebraic formula.

- one equation; $v_t = c_\mu k^{1/2} l$; k from transport equation, l from algebraic formula

- two equation; $v_t = c_\mu k^2/\varepsilon = c_\mu k^{1/2} l$: k and ε (or l) both from transport equations

- stress equation; $\overline{u_i u_j}$ from transport equation

---

The first three approaches of the panel are consistent with the comments of the previous panel and particularly with the assumptions of near equilibrium and high Reynolds number.

The zero equation approach represents a further simplification in that k is not represented in the viscosity formulation. Then, it is implied that the mean motion is unaffected by turbulence intensity and the length scale, l, is presumed known in that it can be specified by an algebraic equation. Prandtl proposed this "mixing-length" approach and it has been used extensively for two-dimensional, boundary-layer flows where

$$v_t = c_\mu l_1^2 \left| \dfrac{dU}{dy} \right|.$$

Experiments indicate that the value of l_1 varies from one free flow to another as is discussed in relation to panels 3.4 and 3.5. It is the simplest approach in that no differential equations, other than those representing conservation of mass and momentum, need be solved. A related algebraic formulation for v_t, due to Cebeci and Smith (1974), is discussed on panel 3.6.

The one-equation approach requires the solution of an equation for turbulence energy and, as a result, allows for its transport and is again consistent with the assumptions of the previous panel. The length scale is specified algebraically and, in common with zero-equation models, this is flow-dependent. The approach has not, as a result, found much favour since it is usually little better than the simpler zero-equation approach. The special form reported by Bradshaw et al. (1967), however, where $-\overline{uv}$ is assumed directly proportional to k, has been extensively used for thin-shear layers with considerable success. An alternative approach reported by Nee and Kovasznay (1969) required the solution of a differential equation with the eddy viscosity as dependent variable. All these formulations are arranged to reduce to the mixing-length formula in the inner region of a wall flow where that formula is valid and where convection and diffusion terms in the differential equations are negligible.

The two-equation approach also accords with the assumptions of the previous panel but is less restrictive, at least in principle, in that both k and l are represented by transport equations. It is clear, however, that the corresponding equations of Chapter 2 cannot be solved without further assumptions which will require to be tested. The eddy viscosity equation of the panel is itself an assumption and is due to Prandtl (1945) and Kolmogorov (1942).

The panel also indicates that a transport equation for $\overline{u_i u_j}$ may, alternatively, be solved. This approach, which will be discussed further in Chapter 4, is not necessarily subject to the restrictions of the previous panel.

3.4 Length scales, 1

	Powers of x for		
	U_s	$y_{1/2}$	$dy_{1/2}/dx$
Plane wake	$-1/2$	$1/2$	
Self-propelled plane wake	$-3/4$	$1/4$	Only approximately
Axisymmetric wake	$-2/3$	$1/4$	self similar
Self-propelled axisymmetric wake	$-4/5$	$1/5$	
Mixing layer		1	0.165*
Plane jet	$-1/2$	1	0.11
Axisymmetric jet	-1	1	0.087
Radial jet	-1	1	0.11

* $U_e = 0$

The above Table, largely borrowed from Tennekes and Lumley (1972), shows that the mean-flow characteristics of jets and wakes, in particular the characteristic velocity U_s and characteristic thickness to the half-velocity point $y_{1/2}$, behave in different ways according to the origins of the flow. It is to be expected, therefore, that although a plane jet may be represented by a particular mixing length formulation ($l_1 \simeq 0.09\ y_{1/2}$), its round counterpart will require a different one ($l_1 \simeq 0.075\ y_{1/2}$). It should be appreciated that the various numbers provided in the panel were obtained from measurements and are subject to some uncertainty. As suggested by Ribeiro and Whitelaw (1980), the growth rate constants for the downstream region of the plane and round jet may not be as different as they appear. For plane jets, the literature shows values from 0.089 to 0.126 and they tend to increase with distance in contrast to the expected self-preservation. A combination of draughts, see Bradshaw (1977b), and jet flapping, see Goldschmidt and Bradshaw (1973), can account for the difference.

In the case of boundary layers, the zero-equation eddy viscosity or mixing length approach has been widely used, for example by Cebeci (see Cebeci and Smith, 1974), for the calculation of boundary layers on airfoils and wings, and has been shown to agree well with related experiments. This is partly due to the relatively mild pressure gradients considered and to the use of an appropriate wall-law formulation. Various Van Driest type approaches (1956) to the near-wall region have been considered, for example, by Jones and Launder (1972), Cebeci et al. (1970) and Pletcher (1970). The eddy viscosity approach of Cebeci and Smith is discussed further in panel 3.6 as a particular example of a zero-equation model. It will be used in the computer programs introduced in Chapter 9.

3.5 Length scales, 2

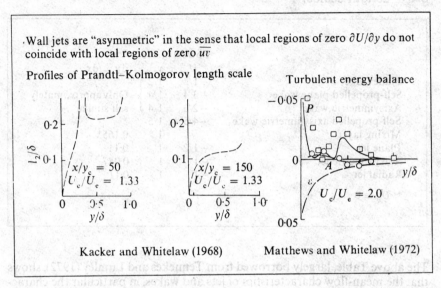

Kacker and Whitelaw (1968)   Matthews and Whitelaw (1972)

The panel presents measured values of the Prandtl–Kolmogorov length scale ($l_2 \equiv \overline{uv}/(\partial U/\partial y)c_\mu k^{1/2}$) obtained downstream of an isothermal wall-jet flow with a velocity ratio of 1·33. As can be seen, the distributions at 50 and 150 slot heights downstream are very different and preclude the use of a simple algebraic formulation for l_2. The results for I_1 (see panel 3.3) allowed a similar conclusion. Thus, for wall flows it is unlikely that a zero or one-equation model

The wall jet has the special property that the mean velocity gradient and local shear stress go to zero at different locations. This property is also found in, for example, annular flows and in many recirculating flows. It implies that an eddy viscosity formulation will lead to infinite viscosity at some locations and zero viscosity at others when both are incorrect. If known, *a priori*, adjustments can be made to a numerical scheme to overcome this difficulty but it indicates a fundamental weakness of the approach. For flows of this type, a stress model is required to represent the detailed flow features.

The panel also shows a turbulence energy balance obtained downstream of a wall jet over a step. Advection (A) is clearly much smaller than production (P) and dissipation (ε) but the ratio of P/ε varies considerably across the layer.

Eddy Viscosity Hypotheses

3.6 Zero equation model of Cebeci and Smith

> - In the inner region, $0 \leq y \leq y_c$, $\quad v_t = l^2 \left|\dfrac{\partial U}{\partial y}\right| \gamma_{tr}$
>
>   with $\qquad\qquad\qquad l = \kappa y [1 - \exp(-y/A)]$
>
> - In the outer region, $y_c \leq y \leq \delta \quad v_t = \alpha \left| \displaystyle\int_0^\infty (U_e - U)\,dy \right| \gamma_{tr}$
>
> $\qquad\qquad\qquad\qquad = 0{\cdot}0168\, U_e \delta^*$ for a wall boundary layer
>
> A, α and γ_{tr} are discussed further below.

Cebeci and Smith (1974) reported an algebraic viscosity formulation for use with external wall boundary layers and this has been used extensively and developed for bodies of revolution, internal flows, three-dimensional flows etc by Cebeci and co-workers. It is based on a Van Driest approach to the inner region, with a damped law of the wall, and on a velocity defect approach to the outer region. The condition which defines y_c is the continuity of the eddy viscosity. The inner-region expression is applied outwards from the wall until the two formulations give the same result; the outer-region expression is used for larger values of y.

The damping constant A may be expressed as

$$A = A^+ \frac{v}{N} U_\tau^{-1}$$

where

$$U_\tau \equiv (\tau_w/\rho)^{1/2}, \qquad N = \left\{ \frac{p^+}{V_w^+} [1 - \exp(11{\cdot}8\, V_w^+)] + \exp(11{\cdot}8\, V_w^+) \right\}$$

and

$$p^+ = -\frac{v U_e}{U_\tau^3}\frac{dU_e}{dx}, \qquad V_w^+ = \frac{V_w}{U_\tau}, \qquad A^+ = 26.$$

For flows with no mass transfer,

$$N = (1 - 11{\cdot}8 p^+)^{1/2}.$$

For values of R_θ less than 5000,

$$\alpha = 0{\cdot}0168 \frac{1{\cdot}55}{1 + \Pi} \quad \text{with} \quad \Pi = 0{\cdot}55 [1 - \exp(-0{\cdot}243 z_1^{1/2} - 0{\cdot}298 z_1)]$$

with $z = (R_\theta/425 - 1)$. For an ordinary boundary layer, with $R_\theta \geq 5000$, the outer-layer expression reduces to that given in the panel.

The intermittency factor, γ_{tr}, accounts for the transition region which exists between the viscous and turbulent flows. For two-dimensional flows, it is defined by:

$$\gamma_{tr} = 1 - \exp\left[-Gx_{tr}(x - x_{tr})\left(\int_{x_{tr}}^{x_{tr}} \frac{1}{U_e} dx\right)\right]$$

where x_{tr} is the location of the onset of transition and the empirical factor G is given by

$$G = \frac{1}{1200} \frac{U_e^3}{\nu^2} R_{x_{tr}}^{-1.34} \quad \text{with} \quad R_{x_{tr}} \equiv U_e x_{tr}/\nu$$

The empirical, but well-tested, nature of this formulation should be noted. Modified versions exist for particular flows and, in all cases, the empiricism implies that the expressions should not be used for configurations far from those upon which the constants are based.

Eddy Viscosity Hypotheses

3.7 Two-equation model

- Solve differential equations with U_i, k and ε as dependent variables
- $v_t = C_\mu k^2/\varepsilon$
- Modelled forms of the k- and ε-equations are required to reduce the number of unknowns to the number of equations
- The result is an approximate representation of turbulent flows: experiments necessary to define range, applicability and precision
- Transport of turbulence energy and dissipation rate is represented but the restrictions of panel 3.2 still apply in principle

A two-equation turbulence model is likely to be more general than simpler alternatives but its quantitative ability to represent a turbulent flow has to be assessed on the basis of a limited range of experiments. In general, the assumptions necessary to render soluble the equations for turbulence energy and dissipation rate cannot be tested directly and assessments have to be made on the basis of calculated dependent variables.

Modelled forms of the k- and ε-equations are considered in the following pages. High Reynolds number is assumed and, for convenience, reference is made to two-dimensional boundary-layer flows. The present form is due largely to Jones and Launder (1972). Alternative forms have been proposed by several authors and some are mentioned in connection with panel 3.9.

The assumption that the Reynolds shear stress is directly proportional to the kinetic energy squared and to the mean velocity gradient, and inversely proportional to the dissipation rate, is more realistic than the assumptions of the zero- and one-equation models in that experiment confirms that k, l and $\partial U/\partial y$ influence $\overline{uv}$. In addition, the solution of equations for k and ε implies that both are calculated, rather than specified.

3.8 Kinetic-energy equation

> - Convection and production terms do not require to be modelled unless normal-stress production is important
> - Diffusion term cannot be retained in exact form: gradient-type diffusion presumed, as in mean-flow equation
> - Viscous diffusion presumed negligible; dissipation may be expressed in terms of $k^{3/2}/l$ in accord with high Reynolds number assumption
>
> $$\frac{Dk}{Dt} = \frac{1}{\rho}\frac{\partial}{\partial x_j}\left(\frac{\mu_t}{\sigma_k}\frac{\partial k}{\partial x_j}\right) + \frac{\mu_t}{\rho}\left(\frac{\partial U_i}{\partial x_j} + \frac{\partial U_j}{\partial x_i}\right)\frac{\partial U_i}{\partial x_j} - \varepsilon$$
>
> where σ_k is an effective Prandtl–Schmidt number

The panel shows a modelled form of the turbulent kinetic-energy equation (panels 1.3, 2.8). The major assumption of gradient diffusion cannot be tested directly since pressure diffusion cannot be measured directly. The two-dimensional boundary-layer form of the unmodelled equation is, in x_1, x_2 coordinates,

$$U_1\frac{\partial k}{\partial x_1} + U_2\frac{\partial k}{\partial x_2} = -(\overline{u_1^2} - \overline{u_2^2})\frac{\partial U_1}{\partial x_1} - \overline{u_1 u_2}\frac{\partial U_1}{\partial x_2} - \frac{\partial}{\partial x_2}\overline{u_2(k + p/\rho)} + \varepsilon = 0$$

and shows the presence of the pressure fluctuation and the triple correlation.

Pressure diffusion is neglected on the evidence of measured turbulence energy budgets which balanced without it; see, for example, Irwin (1973, 1974). The uncertainty in measurements of this type is not insignificant. The arguments for gradient diffusion are not overwhelming and brief discussion has been provided, for example, by Hanjalić and Launder (1972). The modelled equation of the panel contains terms which are known either from the solution of differential equations or from the μ_t-hypothesis.

The normal stress production of the two-dimensional, boundary-layer equation is negligible in many practical situations; in other cases it may be obtained from solutions of normal-stress equations or from the algebraic stress models discussed later in this chapter.

3.9 Dissipation equation

- The dissipation equation contains many terms which are unknown: substantial modelling is necessary

- The exact equation of panel 2.9 may be compared with

$$\frac{\mathrm{D}\varepsilon}{\mathrm{D}t} = \underbrace{\frac{1}{\rho}\frac{\partial}{\partial x_j}\left(\frac{\mu_t}{\sigma_\varepsilon}\frac{\partial \varepsilon}{\partial x_j}\right)}_{\text{Diffusion}} + \underbrace{\frac{C_{\varepsilon 1}\mu_t}{\rho}\frac{\varepsilon}{k}\left(\frac{\partial U_i}{\partial x_j} + \frac{\partial U_j}{\partial x_i}\right)\frac{\partial U_i}{\partial x_j}}_{\text{Production}} - \underbrace{C_{\varepsilon 2}\frac{\varepsilon^2}{k}}_{\text{Dissipation}}$$

Modelled forms of the dissipation equation have been proposed, for example, by Davidov (1961), Harlow and Nakayama (1967), Lumley and Khajeh Nouri (1974) and Hanjalić and Launder (1972). The equation shown on the panel, taken with the eddy viscosity assumption, is representative of these. Only the convection term remains from the exact equation presented in the previous chapter. Of the exact equation, panel 2.9, term (a) has been neglected, (b) and (c) combined to represent production and (e), (f) and (g) combined to represent diffusion.

The equation of the panel, together with the equation of panel 3.8 for the turbulent energy k and the momentum and continuity equations form, with the μ_t assumption, a closed set which can be solved numerically. Alternatively, equations such as Rotta's equation for (kl) can be used to replace that for $(k^{3/2}/l)$. A two-dimensional, boundary-layer form of Rotta's equation (Rotta, 1972b) is

$$U_1\frac{\partial(kl)}{\partial x_1} + U_2\frac{\partial(kl)}{\partial x_2} = \frac{\partial}{\partial x_2}\left[\alpha_{kl}\sqrt{\overline{u_2^2}}l\frac{\partial(kl)}{\partial x_2} + C_{kl}l\frac{\partial k}{\partial x_2}\right]$$

$$- C\overline{u_1 u_2}\frac{\partial U_1}{\partial x_2} - Cl(\overline{u_1^2} - \overline{u_2^2})\frac{\partial U_1}{\partial x_1} - C_s l\varepsilon.$$

Alternative equations have been proposed by Kolmogorov (1942) and Saffman (1974 $\{k/l^2\}$, Saiy (1974) $\{k^{1/2}/l\}$, Shir (1973) $\{l\}$ and Spalding (1972).

3.10 Constants in $k - \varepsilon$ model

C_μ	$C_{\varepsilon 1}$	$C_{\varepsilon 2}$	σ_k	σ_ε
0·09	1.45	1·9	1·0	1·3

- C_μ from law of the wall, stress model and optimization
- $C_{\varepsilon 1}$, $C_{\varepsilon 2}$ decay of grid turbulence and near wall turbulence
- Effective "Prandtl" numbers by optimisation

The coefficients indicated in the panel can be constants in the sense that they are not changed in any one calculation and it is to be hoped that they will not vary much from one flow to another. The values shown are similar to those proposed by Jones and Launder (1972) and are those recommended by Gosman et al. (1979) and Pope and Whitelaw (1976).

The value of C_μ is undoubtedly not constant. The stress model suggests a value of 0·15 and consistency with the law of the wall requires $C_\mu/\sigma_\varepsilon = 0.069$. The decay of isotropic turbulence suggests that $C_{\varepsilon 2}$ has a value of around 1·8 and near-wall turbulence dictates that $(C_{\varepsilon 2} - C_{\varepsilon 1})$ is fixed: values of $C_{\varepsilon 1}$ and $C_{\varepsilon 2}$ are given below:

	$C_{\varepsilon 1}$	$C_{\varepsilon 2}$
Hanjalić and Launder (1972)	1·45	2·0
Wyngaard et al (1974)	1·5	2·0
Launder et al (1975)	1·44	1·9
Gibson and Launder (1976)	1·45	1·9

In curved flows, for example, and as indicated by Rodi (1976) and Gibson (1978), C_μ decreases rapidly with increasing Richardson number and decreases with the ratio of turbulent production to dissipation. This suggests the need for a Reynolds stress model, see Chapter 4, which does not require the specification of a value for C_μ but an algebraic stress model (see panels 3.13 and 3·14) may prove to be adequate for this purpose. The two effective Prandtl numbers are obtained from computer optimization with C_μ/σ_ε taken as 0·069 as mentioned previously.

An extensive range of flows has been calculated with the $k - \varepsilon$ model of turbulence. Some have been reported by Launder and Spalding (1972, 1974). Free flows have been considered by Rodi (1972) and Launder et al. (1973);

Eddy Viscosity Hypotheses

curved flows, with a Richardson number modification, by Priddin (1975) and Launder *et al.* (1977); and recirculating glows by Gosman *et al.* (1979). In general, it has been found suitable for engineering-type calculations though, in detailed terms, it has deficiencies.

3.11 Low Reynolds number flows, 1

- In a number of flow situations, e.g. those with large pressure or density gradients, the law of the wall is invalid and the conservation equations, with low Reynolds number terms may have to be solved
- When $\overline{uv} \not\gg \nu \partial U/\partial y$ the turbulence structure is influenced by fluid viscosity and empirical constants may vary with $(k^2/\nu\varepsilon)$, the turbulence Reynolds number
- There is little experimental data to guide the development of low Reynolds number models

In many wall flows, the near wall region can be adequately represented by the logarithmic law of the wall. This approach, as well as economizing on computer storage and time, ties the mean profile at three locations. In situations where it is inappropriate, for example in laminarizing flows or where adverse pressure gradients are very strong, the conservation equations must be solved to the wall and encompass a region where viscous stresses become increasingly important.

Daly and Harlow (1970) have proposed a stress closure appropriate to flows with low Reynolds number influences; Jones (1971) and Jones and Launder (1972) have made proposals directly relevant to the $k - \varepsilon$ turbulence model; and Hanjalic and Launder (1976) made proposals in connection with their three-equation model. The two-equation approach has been applied, with some success, by Launder and Sharma (1974), Jones and Renz (1974) and Priddin (1975).

3.12 Low Reynolds number flows, 2

$$\frac{Dk}{Dt} = \frac{\partial}{\partial x_i}\left[\left(v + \frac{v_t}{\sigma_k}\right)\frac{\partial k}{\partial x_i}\right] + P - \varepsilon$$

$$\frac{D\varepsilon}{Dt} = \frac{\partial}{\partial x_i}\left[\left(v + \frac{v_t}{\sigma_\varepsilon}\right)\frac{\partial \varepsilon}{\partial x_i}\right] + C_{\varepsilon 1}\frac{\hat{\varepsilon}}{k}P + 2vv_t\left(\frac{\partial^2 U_i}{\partial x_j \partial x_k}\right)^2 - C_{\varepsilon 2}\frac{\hat{\varepsilon}^2}{k}$$

where

$$\overline{u_i u_j} = \tfrac{2}{3}\delta_{ij}k - v_t\left(\frac{\partial U_i}{\partial x_j} + \frac{\partial U_j}{\partial x_i}\right)$$

$$v_t = C_\mu k^2/\varepsilon \quad \text{and} \quad \hat{\varepsilon} = \varepsilon - 2v\left(\frac{\partial \sqrt{k}}{\partial x_j}\right)^2$$

The panel presents equations for turbulence energy and dissipation rate including viscous contributions to diffusion and dissipation. In addition, constants influenced by the viscous regions become functions of turbulence Reynolds number, e.g.

$$C_\mu = 0.9 \exp[-2.51(1 + R_t/50)],$$

$$C_{\varepsilon 2} = 1.9[1 - 0.3 \exp(-R_t^2)].$$

The supposed variation of C_μ is largely pragmatic although that of $C_{\varepsilon 2}$ is based on slightly firmer ground: the latter stems from examination of the weakly anisotropic grid-turbulence data of Comte-Bellot and Corrsin (1966) and recognizing that, although in the near-grid region

$$\frac{D\varepsilon}{Dt} = -C_{\varepsilon 2}\varepsilon^2/k,$$

this formulation has to be altered, as indicated above, in the low Reynolds number region far downstream.

A major deficiency of the above model is that it is likely to fail in the important situation where laminarization is brought about by strong heating. In this case, there will be strong property variations which are not accounted for in the model.

In principle, a universal low Reynolds number formulation should be able to represent transition as well as laminarization. The possibility was explored by Jones who allowed a laminar boundary layer to grow to a predetermined value of Reynolds number at which a perturbation (an energy source) was introduced and allowed to develop or decay. It was found that transition did occur but much further downstream than indicated by experiment.

3.13 Algebraic stress models, 1

- In the Reynolds stress equation (modelled form), the convection and diffusion terms contain gradients of the Reynolds stress: without these terms, the equation is algebraic
- Rodi (1972) assumed

$$\frac{D\overline{u_i u_j}}{Dt} = \frac{\overline{u_i u_j}}{k} \frac{Dk}{Dt}$$

and

$$D(\overline{u_i u_j}) = \frac{\overline{u_i u_j}}{k} D(k)$$

where D represents net diffusion

The two equation model may be modified to take account of stress transport through an algebraic form of the Reynolds stress equation. The assumption used by Rodi (1972) is that the rate of variation $\overline{u_i u_j}/k$ is much less than that of k, and this is often reasonably correct although it is inappropriate to the centre-line region of a jet or wake or to the central region of an annulus flow.

The production, dissipation and pressure strain term in the Reynolds stress equation are thus retained in the forms discussed in the following lecture and the convection and diffusion represented as indicated above. The result is a series of six algebraic equations, representing the components of the stress tensor, which may be solved together with the transport equations of the two-equation model.

Modified approaches have been suggested by Rodi (1976) and Gibson (1978).

3.14 Algebraic stress models, 2

For a two-dimensional thin shear layer, Rodi obtained

$$-\overline{u_1 u_2} = \tfrac{2}{3}(1 - \gamma) \frac{(C - 1 + \gamma P/\varepsilon)}{(C - 1 + P/\varepsilon)^2} \frac{k^2}{\varepsilon} \frac{\partial U_1}{\partial x_2}$$

or

$$v_t = \frac{-\overline{u_1 u_2}}{\partial U_1/\partial x_2} = \tfrac{2}{3}(1 - \gamma) \frac{(C - 1 + \gamma P/\varepsilon)}{(C - 1 + P/\varepsilon)^2} \frac{k^2}{\varepsilon}$$

where γ and C are constants and P represents the production $\overline{u_i u_j}\, \partial U_i/\partial x_j$

The algebraic equation for shear stress deduced by Rodi (1976) may be written as an equation for v_t, which is seen to depend on k^2/ε (same dimensions as v). As P/ε decreases, the coefficient increases and this accords with experimental trends in wakes and in self-propelled bodies when P/ε rapidly decreases. The constants γ and C stem from the two parts of the pressure strain correlation usually written as $\phi_{ij,1}$ and $\phi_{ij,2}$, see Launder (1976). The value of γ is taken to be around 0·6 and C has been assigned values ranging from 0·7 to 5·0.

The algebraic-stress approach was used by Launder and Ying (1973) and by Tatchell (1975) in the calculation of the properties of square-duct flows. Experiments, for example, those of Melling and Whitelaw (1976) have shown that the normal stresses result in secondary motions with maximum velocities slightly greater than 1 per cent of the axial bulk velocity. To represent these secondary flows, which are present in all non-circular duct flows, the normal stresses must influence the calculated results and the algebraic-stress model provides a convenient way of achieving this purpose. Launder and Ying solved a k-equation and assumed an algebraic length-scale formulation. Tatchell, in his later work, used the algebraic-stress model with a two-equation model and achieved reasonable agreement with experiment.

3.15 Scalar fluctuations, 1

- The equations for scalar quantities can be closed at different levels, in the same manner as for momentum
- The simplest approach is to assume a turbulent "Prandtl" number which is specified empirically, i.e. for a two-dimensional flow

$$\frac{\partial}{\partial x}(\rho U \Phi) + \frac{1}{r}\frac{\partial}{\partial y}(r\rho V \Phi) - \frac{\partial}{\partial x}\left(\Gamma_\Phi \frac{\partial \Phi}{\partial x}\right) - \frac{1}{r}\frac{\partial}{\partial x}\left(r\Gamma_\Phi \frac{\partial \Phi}{\partial y}\right) = S_\Phi$$

with $\Gamma_\Phi = \frac{\mu}{\sigma_\Phi}$ and with σ_Φ specified

---

The approach indicated in the panel is still to be recommended for most calculations relating to scalar quantities. For practical heat-transfer calculations, for example, higher-order closures can seldom be justified and the turbulent "Prandtl" number σ_Φ is usually specified as a constant, say 0·9, or simple algebraic function. Measurements of σ_Φ have been reported for a number of boundary-layer flows and appear to range from around 0·5 to 1·5; the measurements are, however, difficult and subject to large errors. Reynolds (1976) has provided a comprehensive review of models of this type.

The measurement of σ_Φ is particularly difficult in the viscous sublayer and the resulting evidence particularly conflicting. Most authors have made use of wall functions corresponding to Van Driest's (1956) mixing length formulations with

$$\frac{U_\tau(\Phi - \Phi_0)}{-\overline{u_2 \phi}} = \sigma_t \left(\frac{U}{U_\tau} + P\right)$$

where P is a factor representing the difference in scalar quantity across the sublayer.

3.16 Scalar fluctuations, 2

- An exact transport equation for $\overline{\phi^2}$ may be obtained by multiplying the equation for $\hat{\phi}$ by 2ϕ and averaging: the result is

$$U_j \frac{\partial \overline{\phi^2}}{\partial x_j} + 2\overline{\phi u_j}\frac{\partial \Phi}{\partial x_j} + \frac{\partial}{\partial x_j}(\overline{u_i \phi^2}) + 2\lambda \overline{\left(\frac{\partial \phi}{\partial x_j}\right)^2} = 0$$

Convection Production Diffusion Dissipation

where λ is the molecular (kinematic) diffusivity of ϕ

- This equation has been modelled as

$$\rho U_j \frac{\partial \overline{\phi^2}}{\partial x_j} - 2\frac{\mu_t}{\sigma_\Phi}\left(\frac{\partial \Phi}{\partial x_j}\right)^2 - \frac{\partial}{\partial x_i}\left(\frac{\mu_t}{\sigma_\Phi}\frac{\partial \overline{\phi^2}}{\partial x_j}\right) + C_{\Phi 2}\frac{\varepsilon}{k}\overline{\phi^2} = 0$$

Corresponding to the turbulence kinetic energy equation, an equation can be obtained for $\overline{\phi^2}$ and is shown on the panel. As expected, it contains triple correlations and dissipation terms which introduce unknowns and render the equation insoluble. A modelled form of this equation which again follows the pattern set by the consideration of turbulent kinetic energy is also presented; it involves the eddy viscosity, gradient diffusion and the assumption of isotropic dissipation. The turbulent "Prandtl" number and dissipation constant $C_{\phi 2}$ require to be specified but the equation does allow the possibility of describing transport of $\overline{\phi^2}$.

Spalding (1971) has used the above modelled equation to represent the concentration fluctuation measurements of Becker et al. (1967). Good agreement was obtained with a value of $C_{\phi 2}$ of 2·0 although 1·2 is required to produce the correct decay behind a grid. A value of 2·0 was also used for the correlation calculations of Chapter 15 with a "Prandtl" number of 0·9.

3.17 Scalar fluctuations, 3

The exact equation of panel 2.11 may be modelled to allow a closure similar to that of the Reynolds stress equation: the assumptions required to obtain the modelled equation are similar for example, for term (III) we have

$$(\lambda + \nu)\overline{\frac{\partial \phi}{\partial x_k}\frac{\partial u_i}{\partial x_k}} = 0 \text{ for isotropic dissipation. Also}$$

$$\overline{\frac{p}{\rho}\frac{\partial \phi}{\partial x_i}} = -C_{\phi 1}\frac{\varepsilon}{k}\overline{u_i\phi} + 0\cdot 8\,\overline{\phi u_m}\frac{\partial U_i}{\partial x_m}$$

$$\qquad\qquad - 0\cdot 2\,\overline{\phi u_m}\frac{\partial U_m}{\partial x_i} + C_{\phi 3}$$

$$-\overline{u_k u_i \phi} = C_\phi \frac{k}{\varepsilon}\left(\overline{u_k u_l}\frac{\partial \overline{u_i \phi}}{\partial x_l} + \overline{u_i u_l}\frac{\partial \overline{u_k \phi}}{\partial x_l}\right)$$

Corresponding to a Reynolds stress closure, the equation for the transport of a scalar flux, see panel 2.11, may be modelled and solved simultaneously with the equations of continuity, momentum, Reynolds stress and enthalpy. The convection and production terms of the equation of panel 2.11 do not need to be modelled and, assuming local isotropy of the fine scale eddies, term (III) may be assumed zero as suggested by Lumley (1972). The pressure-scalar gradient, term (IV), causes most difficulties and has been the subject of considerable research which has been reviewed, in detail, by Launder (1976); the required assumptions are similar to those described in connection with the pressure-strain correlation of the Reynolds stress equation in Chapter 4. Term (V) is represented by gradient diffusion, again in keeping with the assumptions of the Reynolds stress equation.

3.18 Concluding remarks

Chapters 2 and 3 have been concerned with conservation equations in differential form and have, respectively, presented exact and modelled equations. If the exact equations could be solved for general boundary conditions, no assumptions would be required and experiments would be much less in demand. Unfortunately, this is not possible and is unlikely to be possible in the foreseeable future and assumptions are required to render the equations soluble. These assumptions are based on physical reality wherever possible.

The closures considered in this chapter are of the eddy viscosity type and range from algebraic formulations to the frequently used two-equation model. One-equation closures, such as that involving an equation for v_t, have not been discussed since they are unlikely to find much use in the solution of practical problems.

The modelled equations have similar form and can, therefore, be solved by the same numerical procedure although different procedures are required to solve parabolic and elliptic equations. The assumptions concerned with diffusion and dissipation of turbulence quantities cannot be checked directly and those made in the equation representing dissipation rate are not all readily substantiated.

4 Stress Equation Modelling

4.1 Purposes and outline of chapter

- To explain the details of modelling the Reynolds stress transport equations and the associated length scale transport equation
- To discuss the special problems of three-dimensional flow, heat transfer and compressible flow

We begin by showing the relation between the exact Reynolds stress transport equation and the eddy viscosity transport models discussed in the last section, and conclude that the empirical Reynolds stress equation derived from an empirical eddy viscosity transport equation is rather unlike the exact one.

It is necessary to distinguish between methods that can, in principle, represent the complete range of flow geometries without any change in empirical input, and those that need adjustment. There is no good reason to suppose that the former are more reliable in practice.

Models using one or more of the Reynolds stress transport equations need at least one auxiliary equation, for the length scale or a quantity that implies a length scale. The most important unknown term in the Reynolds stress transport equations is the pressure strain "redistribution" term; unfortunately pressure fluctuations within the flow are not measurable at present.

After giving examples of transport equation models we consider some special problems of engineering flows.

4.2 Equivalence of $D\nu_t/Dt$ and Du_iu_j/Dt

Given $-(\overline{u_iu_j} - \tfrac{1}{3}\overline{u^2}\delta_{ij}) = \nu_t(\partial U_i/\partial x_j + \partial U_j/\partial x_i)$
and $D\nu_t/Dt = \ldots$ (e.g. via $\nu_t = c_\mu k^2/\varepsilon$)
then, using $\dfrac{\partial}{\partial x_j}\dfrac{DU_i}{Dt} + \dfrac{\partial}{\partial x_i}\dfrac{DU_j}{Dt} \to \dfrac{D}{Dt}\left(\dfrac{\partial U_i}{\partial x_j} + \dfrac{\partial U_j}{\partial x_i}\right),$

we can deduce equations for $D\overline{u_iu_j}/Dt$. For compatibility with the exact $\overline{u_iu_j}$ equations, the second-derivative terms in the "rate-of-strain transport equation" must cancel with terms in $D\nu_t/Dt$. In existing models they do not, and (in a two-dimensional, thin shear layer)

$$\frac{D(-\overline{uv})}{Dt} = \ldots + \frac{\nu_t}{\rho}\left(\frac{-\partial^2 \overline{P}}{\partial x \partial y} + \frac{\partial^2(-\overline{uv})}{\partial y^2}\right) + \ldots$$

Transport equations for eddy viscosity ensure that the Reynolds stresses follow the mean strain rate. This is an advantage if the Reynolds stresses really do behave like this, and not otherwise. Any eddy viscosity transport model can be converted to a Reynolds stress transport model by using the exact "rate-of-strain transport equation". Usually the resulting Reynolds stress equation is dissimilar to the exact one (see panel 2.7) since it contains $\partial^2 P/\partial x_i \partial x_j$, second derivatives of Reynolds stresses, etc. The main advantage is that eddy viscosity models will always give qualitatively plausible (laminar-like) solutions whereas stress equation models *may* exhibit unsuspected singular behaviour in unfamiliar cases.

4.3 Classification of stress equation models

1. How many $\overline{u_i u_j}$ equations? Shear stress/turbulent energy/all nonzero Reynolds stresses?
2. Length scale from transport equation or algebraic formula?
3. Auxiliary transport equation for other quantities (e.g. triple products?)

Except for corner flows (Chapter 11) and other examples of stress-induced secondary flows, only the shear stress (in local shear-layer axes) need be calculated accurately. It is generally necessary to use shear-layer axes (panel 1.7) for numerical reasons and, at least in the short term, helpful to use the normal to the shear layer as a "preferred direction" in the turbulence model. The thin shear layer equations themselves are non-invariant with respect to rotation of axes because of the use of this preferred direction. Of course if the "preferred direction" can be defined in terms of mean-flow or turbulence properties (Castro and Bradshaw, 1976) the computer can be instructed to redefine its axes internally to conceal the noninvariance from the user. The alternative is a rotationally invariant model that ignores the shear-layer simplifications, calculates all the Reynolds stresses and uses a transport equation to obtain a length scale unrelated to a shear-layer thickness.

Occasionally, transport equations for triple products have been suggested: this implies that the local scales of the energy-containing motion are not sufficient to define the large eddies which dominate the triple products. Various other auxiliary equations have been used for special purposes.

At this level of sophistication, classification of models solely by the number of equations they have is not helpful.

4.4 Strategy of modelling—"complete" methods

- Rotta defines a "complete" set of equations as one that can be applied to any flow geometry without adjustment of empirical constants
- The Navier–Stokes equations form one such set: Is there another (Reynolds averaged or otherwise) that will give good engineering accuracy for all flows of interest?
- Such a "complete" model could be calibrated using data from a wide range of flows

The existence of a *universal* calculation method which is simpler than the numerical solution of the time-dependent Navier–Stokes equations is improbable. Claims of an approach to universality have been made explicitly or implicitly by some developers of calculation methods: for a review see Rodi (1978). Methods that use an eddy viscosity fail to reproduce the detailed behaviour of, for example, asymmetric jet or duct flows where the eddy viscosity deduced from experiments become negative. However some stress equation methods are nominally applicable to any flow geometry, and are realistic enough to cope with regions of negative eddy viscosity so that their universality cannot at once be qualitatively disproved. Use of grid turbulence data (say) for determining the empirical constants in methods intended primarily for shear flows implies belief in a high degree of universality, because grid turbulence results from the interaction of many bluff body wakes.

If a universal method does not exist, we have to employ geometry-dependent empirical constants, and a calculation method cannot be extrapolated to new geometries without new data. Logically speaking, this is a disadvantage only if one would trust the allegedly universal method in new geometries without new data!

In effect, none of the current types of Reynolds stress equation calculation methods can possibly be universal, and faster progress might be made if developers of calculation methods started by optimizing their models for a full range of nonself-preserving boundary-layer flows before applying them to other types of shear layer. Boundary layers subjected to extra rates of strain (panel 1.15) provide a challenging set of test cases. For well-documented and comprehensive catalogues of two-dimensional boundary-layer data see Coles and Hirst (1969: low-speed flow) and Fernholz and Finley (1977: high-speed flow). For a reasoned, cautious view of the universality of a particular calculation method see panels 3.10 and 4.9.

4.5 Energy balance in mixing layer

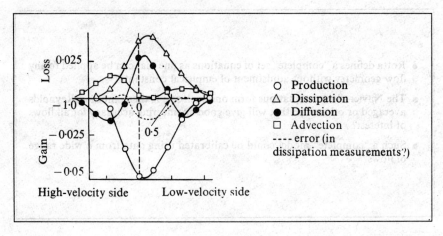

These measurements of the terms in the turbulent energy equation in a mixing layer (panel 1.8) are from Castro (1973: see Castro and Bradshaw, 1976). Production peaks quite sharply in mid-layer, as does dissipation, but near the edges the transport of turbulent energy by the mean flow (advection) is nearly balanced by lateral transport by the turbulence (diffusion), and production and dissipation are much smaller. Note that maximum production is roughly twice maximum dissipation; local-equilibrium arguments (panel 1.9) do not apply to free shear layers.

A boundary-layer energy balance looks roughly like one half of the above but with production and dissipation nearly equal for $y/\delta < 0.2$ and larger than advection or diffusion for $y/\delta < 0.8 - 0.9$. Local-equilibrium arguments (panel 1.9) apply for $y/\delta < 0.1 - 0.2$, diffusion and advection being negligible.

4.6 The "pressure-strain" term

- From $\dfrac{\partial}{\partial x_i}\left(\dfrac{\mathrm{d}\hat{U}_i}{\mathrm{d}t}\right)$ we get, after subtracting the mean part,

$$-\frac{1}{\rho}\frac{\partial^2 p}{\partial x_i^2} = \frac{\partial U_i}{\partial x_j}\frac{\partial u_j}{\partial x_i} + \frac{\partial^2}{\partial x_i \partial x_j}(u_i u_j - \overline{u_i u_j}) \equiv R,\text{ say}$$

- The solution is

$$\frac{p}{\rho}(\mathbf{x}) = -\frac{1}{2\pi}\iiint \frac{R(\mathbf{x}^*)}{|\mathbf{x}-\mathbf{x}^*|}\,\mathrm{d}V(\mathbf{x}^*)$$

- This shows that p, and $\overline{p(\partial u_l/\partial x_m + \partial U_m/\partial x_l)}$, depend on the weighted integral of $\partial U_i/\partial x_j$ over an eddy correlation volume

It is surprising at first sight that pressure fluctuations depend on the mean strain rates as well as on velocity fluctuations. Physically—and crudely!—pressure fluctuations can be set up either when one eddy encounters another *or* when an eddy encounters a region with a mean velocity different from its own.

The pressure fluctuation at a point depends on an integral of the right-hand side R of the Laplace equation over the whole of the flow. The pressure-strain term (written here with indices l and m to avoid confusion) depends on the integral of the product of R at $\mathbf{x}^*$ with the strain-rate fluctuation, $(\partial u_l/\partial x_m + \partial u_m/\partial x_l)$, at $\mathbf{x}$. For values of $|\mathbf{x}-\mathbf{x}^*|$ greater than the largest eddy length scale, the product is negligible.

If a solid surface is present the integral must extend over the exact image flow under the surface, as well. If y is not large compared with a large eddy scale (and it never is in an attached boundary layer) the image contribution is significant.

Use of local $\partial U/\partial y$ in a model for the pressure-strain term is a gross approximation. For an illuminating treatment of the homogeneous case ($\partial U_i/\partial x_j$ constant everywhere) see Crow (1968). A few calculation methods survive in which the contribution of $\partial U_i/\partial x_j$ to the pressure-strain term is neglected, and only the contribution of the second half of R retained. By suitable adjustment of constants such methods can give correct answers in local equilibrium flows but it is hard to see how they can perform well in cases where the mean strain rate and the turbulence are not in equilibrium.

At present there is no accepted way of measuring pressure fluctuations. Probably the main source of information in the next few years will be large-eddy simulations (solutions of the time-dependent Navier–Stokes equations for the energy-containing motions, with the small-scale eddies represented by some turbulence models: see Ferziger, 1977). Since the pressure-strain term in the shear-stress transport equation is a large one it can be measured by difference, although this gives no detailed information.

In general the other terms in the individual Reynolds stress transport equations are modelled in much the same way as the terms in the turbulent energy equation (from which the pressure-strain term cancels out). At least some of the generation terms are usually kept in exact form.

Stress Equation Modelling

4.7 $\overline{uv}$ balance in mixing layer

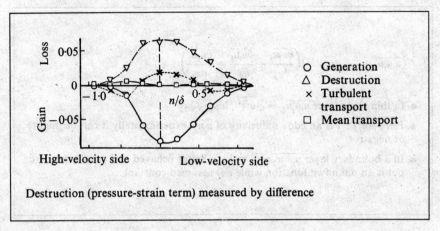

Destruction (pressure-strain term) measured by difference

The data are again from Castro. At first sight the mean and turbulent transport terms are less important, relative to the generation/destruction terms, than in the case of the turbulent energy balance. However Reynolds averaging, with the pressure left explicit rather than represented by the "solution" of the equation in panel 4.6, groups the mean strain dependent part of the pressure-strain term with the turbulence-dependent part. It would be more realistic to subtract the former part from the (mean-strain-dependent) generation term, because pressure fluctuations directly oppose the generation of Reynolds stress. With this grouping the shear-stress balance would look more like the energy balance—as it should, since the ratio of shear stress to turbulent energy is constant over most of the flow.

4.8 Modelling of turbulent transport terms

$$\overline{u_i u_j u_k} = -c(\overline{u_i^2})^{1/2} l \left(\frac{\partial \overline{u_i u_j}}{\partial x_k} + \frac{\partial \overline{u_i u_k}}{\partial x_j} + \frac{\partial \overline{u_j u_k}}{\partial x_i} \right)$$

- In thin shear layer $\overline{u_i u_j u_2} = c(\overline{u^2})^{1/2} l \partial \overline{u_i u_j}/\partial x_2$
- Here $c(\overline{u_i^2})^{1/2} l$ is an eddy diffusivity of $\overline{u_i u_j}$: experimentally it can be infinite or negative.
- In a boundary layer $\overline{u_i u_j u_2}/\overline{u_i u_j} \equiv V_{ij}$ is better behaved and usually positive, but is an unknown function while c is assumed constant

The customary modelling of triple product terms is by a gradient-diffusion form of the type shown above. This form is rotationally invariant: in a thin shear layer the only non-negligible gradient is $\partial/\partial x_2$. Turbulent transport by pressure fluctuations is lumped in with the triple products on the argument that it is part of the same process. Pressure-driven transport of turbulent energy is small near the outer edge of a shear layer; production and dissipation are negligible and mean transport (advection) seems to be closely balanced by triple-product "diffusion". This is insufficient evidence for assuming pressure-driven transport of each $\overline{u_i u_j}$ to be zero everywhere. Pressure-driven transport of turbulent energy includes energy loss by sound radiation but this is negligible at nonhypersonic speeds.

Gradient diffusion is a good model of momentum or heat transport by molecules in a dense gas (mean free path ≪ flow dimensions). The "mean free path" of eddies, however defined, is *not* small compared to the width of a shear layer, so the gradient diffusion model can be justified only as a vehicle for data correlations, based on the expectation that turbulent transport of $\overline{u_i u_j}$ will generally be down the gradient of $\overline{u_i u_j}$ if not really proportional to it. However the general rule in boundary layers seems to be that triple-product transport is outwards from the surface, whatever the direction of the gradient. In a curvature-destabilized boundary layer the shear stress $-\overline{uv}$ reaches a high peak in midlayer (Smits *et al.*, 1979), but the y-component flux $-\overline{uv^2}$ is positive everywhere, implying a negative diffusivity in the inner part of the layer.

An alternative approach is to define a turbulent transport velocity in the direction normal to the shear layer like V_{ij} above or V_t, V_L in panel 4.11. For a given type of shear layer it should be close to a universal function of y/δ multiplied by a velocity scale of the large eddies. The latter may not be easy to relate to the Reynolds stresses. Transport equations for triple products or at least for the "large eddy velocity scale" may be needed in some cases.

Of course eddy diffusivity and turbulent transport velocity can be related by equating the two expressions for the triple product. The question is which representation is the more suitable, on grounds of physical plausibility or computational convenience.

East and Sawyer (1979) have found, in measurements of self-preserving ("equilibrium") boundary layers in pressure gradients, that the y-component flux of turbulent shear stress, $-\overline{uv^2}$, changed sign where $\partial(-\overline{uv})/\partial y$ changed sign, in contrast to the behaviour of the curved shear layer mentioned above. Indeed, they showed that the eddy diffusivity of shear stress in their experiments was well represented by the simple expression $3l(-\overline{uv})^{1/2}$ where l is the mixing length.

4.9 Transport equations for length scale

> - Problems are the same in Reynolds stress transport modelling as in eddy viscosity transport modelling (if length scales meaningful)
> - Since no terms in exact length-scale equations have been measured, why not go straight to a physically plausible empirical equation?
> - Equations for $V^m L^n$ can always be converted into equations for L and their merits examined. $\varepsilon \equiv V^3/L$ has the merit that its turbulent transport is fairly well described by an eddy diffusivity

Explicit consideration of a length scale, rather than a quantity like dissipation rate ε from which a length scale can be derived, is helpful conceptually. In a shear layer oriented model the influence of the shear-layer thickness on the eddy length scale can be included explicitly (but this is open to the same objections as the use of eddy viscosity, that it amounts to keeping one foot on the bottom when venturing into deep water).

It can be argued that none of the exact transport equations for length scales, or quantities yielding length scales, is sufficiently well documented experimentally to be a useful constraint on modelling. According to this viewpoint one might as well use a wholly empirical equation containing dimensionally correct terms simulating the usual transport equation processes. Exactly the same considerations apply to the length-scale equation used in two-equation eddy viscosity transport models. For a given definition of length scale, the empirical transport equation should of course be the same in an eddy viscosity transport method as in a stress equation method. Therefore experience gained with the former methods should be applicable to the latter.

4.10 Response of length scale to extra strain rates

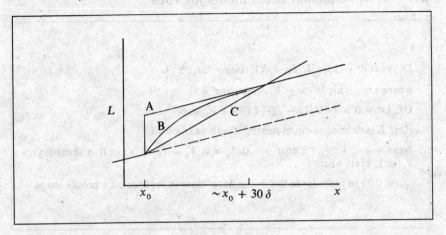

The effect of a sudden application of extra strain rate (e.g. a sudden change in surface curvature at $x = x_0$, say) takes some time to affect the turbulence structure. That is, the response of L, say, will be like curve B not curve A. The response predicted by a DL/Dt equation with a term on the right-hand side proportional to the extra strain rate is like curve C—the slope being disposable—and the behaviour of the right-hand side term required to reproduce curve B is complicated.

What is lacking is the mechanism by which L/δ in the real flow achieves a new equilibrium level, $(L/\delta)_{eq}$. It would be very difficult to define $(L/\delta)_{eq}$ beforehand in any model intended for wide application, so the addition of terms like $c[(L/\delta)_{eq}]\tau^{1/2}$ to the DL/Dt equation would be too trivial. This is the point at which one has to ask whether the length scale(s) required for modelling the Reynolds stress equations necessarily behave like real eddy length scales such as the correlation integral scale (L_x in Panel 1.5). It seems likely that they do not.

There seems to be a need for equations for dimensionless structure parameters to act as "disequilibrium indices". Some current models which solve for all the Reynolds stresses use Reynolds stress ratios (anisotropy ratios) in this way but these are linked only indirectly with the higher-order quantities like length scale, triple products and pressure strain terms.

4.11 A two-equation stress model for TSLs

$D(-\overline{uv})/dt = 2a_1\tau \partial U/\partial y - \partial(V_\tau \tau)/\partial y - 2a_1 \tau^{3/2}/L$

where $\tau \equiv -\overline{uv}$, $V_\tau \equiv -V_{12}$ (see panel 4.8)

$DL/Dt = 0 \times L\partial U/\partial y - \partial(V_L L)/\partial y + c_{SL}\tau^{1/2}$

Here L is defined to equal mixing length near a wall

Near $y = \delta$, $V_L \sim V_\tau$; near $y = 0$, $V_\tau \approx 0$, $V_L = c_{SL}U_\tau/\kappa$: in free shear layers $V_L \approx V_\tau$ everywhere

$c_{SL} = 0.05$ in inner layer but in the outer layer it depends on τ profile shape

---

The simplest possible stress equation model, like the simplest eddy viscosity transport model, has two equations; one is for the velocity scale, necessarily $(-\overline{uv})^{1/2}$, and the other for the length scale. The original model of Bradshaw et al., (1967) used an algebraic equation for length scale; this, with an empirical correction for the effects of extra strain rate (panel 1.16) which included a crude transport equation, has served well in a wide variety of flows. A length-scale transport equation has now been developed, and tested with a wide variety of the Stanford conference test cases (Coles and Hirst, 1969). Two important qualitative conclusions are (i) that gradient transport (gradient diffusion) is not suitable for a length-scale equation (it is easy to see that the diffusivity becomes negative near a solid surface) and (ii) that the "constants" in a simple length-scale equation like that above vary with the shape of the shear stress profile, or possibly with $d\delta/dx$ as such. The choice of ε or (velocity scale)^3/(length scale) as a variable seems to permit use of the gradient diffusion model for turbulent transport but seems unlikely to make all the constants in the equation truly constant. However most model equations for length-scale quantities show, when rearranged to give DL/Dt explicitly, that the term in $\partial U/\partial y$ is small. Since the mixing length formula $L\partial U/\partial y = \tau^{1/2}$ is rarely wrong by an order of magnitude, small $\partial U/\partial y$ terms could be hidden in c_{SL}.

In the form used by Bradshaw and Unsworth (1977) the extra strain rate corrections in this model are the same as those used in its algebraic-L predecessor. L is calculated from the equation shown here and then factored by a correction derived from a simple lag equation to give a response like curve B in panel 4.10. Note that V_τ in the present notation is $2a_1 \tau_{max}^{1/2} G(y/\delta)$ in the notation of Bradshaw et al.

4.12 A full stress equation model

> The two-dimensional, thin, shear-layer version of Hanjalić and Launder's model uses $D\overline{uv}/Dt$ and $D(\overline{u_i^2}/2)Dt$ equations qualitatively like panel 4.10 (with gradient diffusion for turbulent transport terms) and
>
> $$\frac{D\varepsilon}{Dt} = -1.45\,\varepsilon\,\frac{\overline{uv}}{k}\frac{\partial U}{\partial y} - \frac{1.9\,\varepsilon^2}{k} + 0.075\frac{\partial}{\partial y}\left(\frac{k^2}{\varepsilon}\frac{\partial\varepsilon}{\partial y}\right)$$
>
> where $k = \tfrac{1}{2}(\overline{u^2} + \overline{v^2} + \overline{w^2})$
>
> Here $k^{3/2}/\varepsilon$ is used as a length scale, about 6 times the mixing length near a wall. The general method solves for ε and all $\overline{u_i u_j}$

Hanjalić and Launder (1972: for the full development see Launder et al., 1975, and Launder and Morse, 1979) produced one of the first methods in which a length scale and all non-zero Reynolds stresses are calculated. This is a necessary requirement for a calculation method which is to be independent of flow geometry and coordinate axes. They make extensive use of rotational-invariance conditions to simplify the modelling of the pressure-strain terms in the Reynolds stress transport equations. The turbulent energy dissipation rate ε was chosen as the length-scale variable, and gradient-diffusion assumptions were made for all turbulent transport terms.

They assumed that (at high Reynolds number) the coefficients in their equations were absolute constants, the same in all kinds of turbulent flow. Some of the constants were evaluated from the decay of grid turbulence, others from shear-layer measurements. At present the results for plane wall jets and circular free jets are poor, and there appear to have been no comparisons with boundary layer data other than in zero pressure gradient. Nevertheless the results are promising and this general approach is certainly the right one for a method intended to be applied to a wide range of flows: probably at least some empirical adjustment of the constants from flow to flow will be needed. For a noninvariant version of this model, see panel 4.16.

4.13 Three-dimensional thin shear layers

$$\frac{D}{Dt}(-\overline{uv}) = \ldots - \overline{p\left(\frac{\partial u}{\partial y} + \frac{\partial v}{\partial x}\right)} + \ldots (y \text{ perpendicular to layer})$$

$$\frac{D}{Dt}(-\overline{vw}) = \ldots - \overline{p\left(\frac{\partial w}{\partial y} + \frac{\partial v}{\partial z}\right)} + \ldots$$

- In two-dimensional flow, $\overline{p\left(\frac{\partial u}{\partial y} + \frac{\partial v}{\partial x}\right)} = \alpha \frac{\partial U}{\partial y} + \gamma$, where α, γ are turbulence turbulence quantities

- In three-dimensional flow, does $\overline{p\left(\frac{\partial u}{\partial y} + \frac{\partial v}{\partial x}\right)} = \alpha \frac{\gamma U}{\partial y} + \beta \frac{\partial W}{\partial y} + \gamma$?

Turbulence is always instantaneously three-dimensional, so that three-dimensionality of the mean flow should change the statistical symmetries without much change in the turbulence structure. To a first approximation the modelling of the $D\overline{uv}/Dt$ equation should therefore be unchanged. However Rotta (1976) suggests that the all-important pressure-strain term $\overline{p(\partial u/\partial y + \partial v/\partial x)}$ depends on $\partial W/\partial y$ in a sense that explains the anisotropy of eddy viscosity observed in many experiments, and the way in which most calculation methods overestimate the resistance of three-dimensional boundary layers to separation (van den Berg et al., 1975).

Rotta's suggestion that the coefficient of $\partial W/\partial y$ depends on W/U violates the Galilean invariance requirement that the turbulence should not depend on the velocity of the observer, but most other obvious choices violate the requirement of invariance with respect to rotation about the y axis. Most existing three-dimensional boundary-layer data follow the inviscid secondary flow formula $W/U = -(\partial W/\partial y)/(\partial U/\partial y)$ fairly closely in the outer layer, so Rotta's suggestion could be rescued by this substitution and applied whether the formula applied or not. The physical explanation may be a history effect, the pressure strain term depending on the direction of the mean strain rate some distance upstream: in that case $\partial(\partial W/\partial y)/\partial x$, rather than $\partial W/\partial y$, should enter.

Schneider (1977) has tested Rotta's suggestion; Rotta (1979) has produced a new, Galilean-invariant model involving $\partial^2 W/\partial y^2$, but although more rigorous it seems to be less successful than the 1976 model.

Three-dimensional slender shear layers—e.g. flows in noncircular ducts—are discussed in Chapter 11.

4.14 Heat or scalar–pollutant transfer

> $D\overline{u_i u_j}/Dt = \ldots + \overline{p(\partial u_i/\partial x_j + \partial u_j/\partial x_i)} + \ldots$
>
> $D\overline{u_i T'}/Dt = \ldots + \overline{p \partial c T'/\partial x_i}$
>
> noting that $\overline{p \partial T'/\partial x_i} = -\overline{T' \partial p/\partial x_i} + \overline{\partial p T'/\partial x_i}$
>
> - In a thin shear layer, $\overline{p \dfrac{\partial T'}{\partial y}} \approx -c_1 \dfrac{2\varepsilon}{q^2} \overline{vT'} + c_2 \overline{vT'} \dfrac{\partial U}{\partial y}$
>
>   (or possibly $\overline{uT'}$ instead of $\overline{vT'}$ in the second term)
>
> - As shown, $c_{1T} \simeq 4$, $c_{2T} \simeq 0.5$ (Launder) and the first term is roughly three times the second
>
> - Most of the problems in the $\overline{u_i u_j}$ equation appear again

The enthalpy or species conservation equations do not contain pressure fluctuations, but the transport equation for the turbulent heat transfer rate in the x_i direction, $\rho c_p \overline{T' u_i}$, contains $\overline{p \partial T'/\partial x_i}$ or $\overline{T' \partial p/\partial x_i}$ and this term is the most difficult to model. Launder (1976) discusses it in detail. Because the mean strain rate appears in the p equation, it must appear in any model for $\overline{p \partial T'/\partial x_i}$. As in the case of the pressure–strain term in the $\overline{u_i u_j}$ equation the model must also contain a term composed entirely of turbulence quantities. There is *no* term depending on the mean temperature gradient. Launder also considers the contribution of buoyancy forces to $\overline{p \partial T'/\partial x_i}$.

There are no measurements of $\overline{p \partial T'/\partial x_i}$, and few of any of the other terms in the $D \overline{u_i T'}/Dt$ equation. Modelling of turbulent transport of $\overline{u_i T'_i}$ usually employs gradient diffusion, and in general the modelling of $D \overline{u_i T'}/Dt$ is closely analogous to that of $D \overline{u_i u_j}/Dt$. Too close an analogy would lead to prediction of simple relations between heat transfer and momentum transfer, such as a simply behaved turbulent Prandtl number σ_t. For many purposes the latter may suffice—the most important source of error in heat-transfer predictions is error in calculating the velocity field. However, transport-equation modelling by Launder (1976), Lumley (1979) and others is helping to establish the limits of simple assumptions for σ_t. Townsend (1976) and Launder have pointed out that σ_t is smaller near the edge of a shear layer than near a solid surface. They offer two plausible but completely different explanations.

In buoyant flows there is an interaction between the density fluctuation field and the velocity fluctuation field. Much of the advanced work in turbulence modelling and large eddy simulations is related to geophysical problems (see for example Wyngaard *et al.*, 1974). A short review of buoyancy problems is given by Bradshaw and Woods (Bradshaw, 1976, Chapter 4).

4.15 Compressible flow

- In conventional variables without mass averaging,
  shear stress $= -\bar{\rho}\,\overline{uv} - \overline{\rho'uv}$,
  If $\rho'/\bar{\rho} \approx (\gamma - 1)M^2 u/U$, then
  $\overline{\rho'uv}/\overline{\rho uv} \approx (\gamma - 1)M^2 \overline{u^2 v}/(U\overline{uv})$
  $\leqslant 0{\cdot}01\, M^2$ (roughly) in boundary layers
  $\leqslant 0{\cdot}05\, M^2$ in mixing layers
- Morkovin's hypothesis does not cover mean density gradients

The explicit density-dependent terms are either easy to deal with—for instance $\overline{\rho'v}$ always appears with ρV and we can define a new variable $\hat{\rho}\hat{V} = \bar{\rho}V + \overline{\rho'v}$—or small as in the case above. This applies in qualitatively the same way to high Mach number flows and to low-speed flows with large density differences. In both cases an equation for density (i.e. temperature or species concentration) must be solved, but a turbulent Prandtl number assumption should suffice. With the very large density variations found in combusting flows (see Chapter 15) the solution of equations for density-velocity and density scalar quantities may be required.

Morkovin's hypothesis, that ρ' has a small effect on turbulence structure if $\rho'/\bar{\rho} \ll 1$, authorizes the use of incompressible models for $M < 5$ in a boundary layer, $M \leqslant 1$ in a mixing layer, if *mean* density gradients have a small effect. Surprisingly, $\partial\bar{\rho}/\partial y$ seems to have little effect even on the large eddies. The good results given by the mixing-length formula in constant-pressure nonhypersonic boundary layers confirms the insensitivity of the turbulence to ρ' and $\partial\bar{\rho}/\partial y$. However $\partial\bar{\rho}/\partial x$ can have surprising effects: in adverse pressure gradients, the local c_f, $\tau_w/(\tfrac{1}{2}\rho_e U_e^2)$, in a supersonic boundary layer can *rise*, instead of falling as low-speed precedent and consideration of the effect of pressure gradient on profile shape would suggest. The recent compressible-flow methods of Wilcox and of Donaldson seem to reproduce this effect (Rubesin et al., 1977) but it is not clear how. Narasimha and Viswanath (1975) show that expansion reduces turbulence intensity very considerably, although their experiment contained stabilizing curvature as well as expansion.

4.16 Present state of $\overline{u_i u_j}$ models

> Models solving for all non-zero $\overline{u_i u_j}$ and a length scale quantity include those of Launder *et al.*, Donaldson, and Lumley. Range of applicability is still not clear—performance in self-preserving flows is often good but there seems to have been few comparisons with flows that are not self-preserving

It would not be wise to attempt a comparison of this year's models, since rapid development continues. Launder and Morse (1979) discuss the shortcomings of Reynolds stress transport models, and Hanjalić and Launder (1979) present an empirical modification to their dissipation equation to allow for effects of normal stresses. The resulting equation is not rotationally invariant, but this is not a fatal drawback if shear-layer axes are used. See also the comparison by Rubesin *et al.* (1977) of the performance in compressible boundary layers of Donaldson's stress equation model and the eddy viscosity transport model of Wilcox (originally due to Saffman 1974).

A surprising feature of the transport equation modelling scene is the failure of developers to test their models over the rather wide range of flows that are not self-preserving for which data are now available. An exception is the paper by Gibson (1979) whose calculations qualitatively reproduce the anomalous oscillatory behaviour of the Reynolds stresses in the curved mixing layer of Castro and Bradshaw (1976). For the present, anyone using *any* calculation method is strongly recommended to test it against data for flows similar to those in which the method is to be used. Where such data are not available no guarantees can be given.

4.17 Concluding remarks

Key questions in modelling include:

> To what extent are the "constants" the same in different flows?
> Is the use of local strain rate in p adequate?
> Can a finite length scale be modelled in terms of quantities at a single point?
> How are the constants affected by wall proximity?
> Is gradient diffusion an adequate model of turbulent transport?
> Can the effects of extra strain rates be modelled without extra data?

The overall message of these concluding remarks is given at the end of the last page; models can be excellent as vehicles for the interpolation of measurements but extrapolation must be considered with great care. Models are based on assumptions and involve empirically based numbers which may be considered as sensibly constant over a particular range of flows. "Universal constants" do not exist. The remaining five questions posed above will have answers which are also flow dependent—and measurements are necessary to provide answers.

5 Introduction to Numerical Methods for the Solution of the Thin Shear-layer Equations

5.1 Purposes and outline of chapter

- To introduce basic finite-difference methods for thin shear-layer problems
- To discuss initial and boundary conditions
- To describe and compare two popular implicit finite-difference methods, namely that of Crank and Nicholson and Keller's Box scheme

The chapter begins with a classification of the forms of equation appropriate to fluid mechanic problems and an indication of the corresponding solution methods. This chapter, together with Chapters 6, 7, 8 and 9, is concerned with boundary-layer flows to which the thin shear-layer equations are appropriate. These equations are parabolic in form and have specific boundary and initial condition requirements which are discussed.

A general introduction to finite-difference solution methods is provided. Implicit and explicit schemes are identified and two popular implicit schemes described. The Crank–Nicholson method has provided the basis for several boundary-layer calculations and Keller's Box method has proved to be particularly useful for the solution of parabolic partial differential equations. The Box method will be used in connection with the solutions of Chapters 6 and 9, with the stability and transition problems of Chapter 12 and with the wing calculations of Chapter 13.

5.2 Classification of partial differential equations

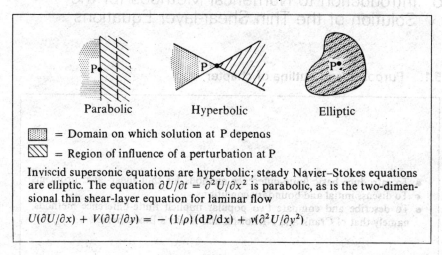

Parabolic   Hyperbolic   Elliptic

▓ = Domain on which solution at P depends
▧ = Region of influence of a perturbation at P

Inviscid supersonic equations are hyperbolic; steady Navier–Stokes equations are elliptic. The equation $\partial U/\partial t = \partial^2 U/\partial x^2$ is parabolic, as is the two-dimensional thin shear-layer equation for laminar flow

$$U(\partial U/\partial x) + V(\partial U/\partial y) = -(1/\rho)(dP/dx) + \nu(\partial^2 U/\partial y^2)$$

The names "elliptic," "parabolic," and "hyperbolic" are attached to partial differential equations according to the geometry of their domains of dependence and regions of influence as shown on the panel. If we imagine that the sketches above refer to fluid-flow problems with flow from left to right, a perturbation at the point P has no upstream influence if the relevant equations are parabolic or hyperbolic, but does have upstream influence if the equations are elliptic. The steady Navier–Stokes equations are elliptic but the thin shear-layer or "boundary-layer" approximations lead to parabolic equations. In the thin shear-layer equations, upstream influence has been suppressed, which means that a numerical solution can be obtained by "marching" in the downstream direction: given the initial conditions on the vertical line through P in the sketch, we can use the differential equation to predict conditions on the dotted line one step further downstream, and so proceed as far downstream as we wish in one sweep by taking many steps in turn. In the case of elliptic equations, a solution scheme of this sort would have to be repeated for many iterations so that upstream influence can be gradually propagated. Hyperbolic equations can be solved by one marching "sweep"; their domain of dependence will be discussed further when we deal with steady three-dimensional thin shear-layer equations in Chapters 8 and 13.

Elliptic equations like the Navier–Stokes equations are much more expensive to solve than parabolic equations like the thin shear-layer equations. In Chapters 10 and 11 we shall discuss cases in which elliptic equations, or so-called "partially parabolic" equations admitting *some* upstream influence, have to be solved. In this chapter we consider the simpler problem of marching solutions for parabolic equations, especially the thin shear-layer equations.

For a further discussion of the classification of partial differential equations and their domains of dependence, see Garabedian (1964), Isaacson and Keller (1966), and Cebeci and Bradshaw (1977).

5.3 Numerical solution of partial differential equations

- The solution of the equation

$$\frac{\partial U}{\partial t} = \frac{\partial^2 U}{\partial x^2} \qquad 0 \leqslant x \leqslant L$$

can be obtained by four different methods:
  1. Finite-difference methods
  2. Method of lines
  3. Galerkin method
  4. Finite-element methods

- Solution of the equation requires:
  a. Boundary conditions,
     i.e. at $x = 0$, U = given, $x = L$, U = given
  b. Initial conditions
     at $t = 0$, $U = U(x)$ given

In general, the four numerical methods of the panel may be used to solve partial differential equations. A brief description of these methods has been provided by Cebeci and Bradshaw (1977) in the context of fluid mechanics problems; Isaacson and Keller (1966) provide a more detailed description. Finite difference methods have proved to be the most successful, especially for the solution of the thin-layer equations, and the present discussion is concerned only with them.

In this introductory chapter finite difference methods are described in relation to the equation of the panel, which has the form of the unsteady, one-dimensional heat-conduction equation. The solutions will march in the t-direction. Note that the ^ has been omitted from the symbol U for simplicity of notation even though $U = U(x, t)$.

No problem statement is complete without the specification of boundary and initial conditions; with the above equation these correspond to the velocity (or velocity gradient) at the spatial boundaries $x = 0$ and $x = L$ and at the time $t = 0$. Boundary and initial condition requirements are discussed in this chapter, particularly in relation to panels 5.16 and 5.19.

5.4 Finite difference approximations to derivatives, 1

> By Taylor's theorem we can write
> $$U(x + h) = U(x) + hU'(x) + \tfrac{1}{2}h^2 U''(x) + \tfrac{1}{6}h^3 U'''(x) + \ldots \qquad (1)$$
> and
> $$U(x - h) = U(x) - hU'(x) + \tfrac{1}{2}h^2 U''(x) - \tfrac{1}{6}h^3 U'''(x) + \ldots \qquad (2)$$
> Add (1) and (2) to get
> $$U(x + h) + U(x - h) = 2U(x) + h^2 U''(x) + O(h^4) \qquad (3)$$
> Here $O(h^4)$ denotes terms containing fourth and higher powers of h. Solving for $U''(x)$ we get
> $$U''(x) = \frac{1}{h^2}\left[U(x + h) - 2U(x) + U(x - h)\right] \qquad (4)$$
> with a leading error on the right-hand side of $O(h^2)$

This and the following page are concerned with the approximation of derivatives. Taylor's theorem applied to the velocity U over an increment h in the x-direction provides the expansions of equations (1) and (2), which have been truncated at the terms with third derivatives. The use of $\pm h$ corresponds to central differencing. The addition of equations (1) and (2) leads to equation (3) which may be inverted to give the second derivative in the form of equation (4).

Equation (4) involves an approximation which stems from the truncation of the Taylor expansion. As $h \to 0$ the error will tend to zero and will in fact be of order h^2 provided that the velocity distribution is sufficiently smooth (for example, four times continuously differentiable).

5.5 Finite difference approximations to derivatives, 2

Subtract (2) from (1)
$$U(x + h) - U(x - h) = 2hU'(x) + O(h^3),$$
which yields
Central Difference Formula, error of $O(h^2)$,

$$U'(x) = \frac{1}{2h}[U(x + h) - U(x - h)] \qquad (5)$$

Forward Difference Formula, error of $O(h)$,

$$U'(x) = \frac{1}{h}[U(x + h) - U(x)] \qquad (6)$$

Backward Difference Formula, error of $O(h)$,

$$U'(x) = \frac{1}{h}[U(x) - U(x - h)] \qquad (7)$$

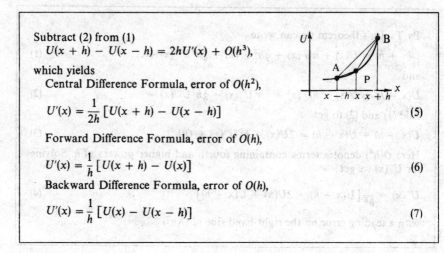

The panel contrasts the expression for the first derivative obtained by considering central, forward and backward differences. As can be seen from the figure, equation (5) approximates the derivative at P by the slope of the chord AB and corresponds to a central difference formulation. We can also approximate this derivative by the slopes of the chords PB and AP, corresponding to forward and backward differences, respectively. It is easy to see that the chord AB is a more accurate representation, with an error of order h^2 rather than the error of order h which arises in the forward and backward difference formulations. Equations (6) and (7) can also be obtained directly from equations (1) and (2) if the expansion is truncated at the first derivative.

5.6 Finite difference notation for functions of several variables

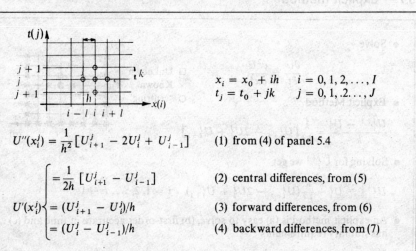

$$x_i = x_0 + ih \quad i = 0, 1, 2, \ldots, I$$
$$t_j = t_0 + jk \quad j = 0, 1, .2\ldots, J$$

$$U''(x_i^j) = \frac{1}{h^2}[U_{i+1}^j - 2U_i^j + U_{i-1}^j] \qquad \text{(1) from (4) of panel 5.4}$$

$$U'(x_i^j) \begin{cases} = \dfrac{1}{2h}[U_{i+1}^j - U_{i-1}^j] & \text{(2) central differences, from (5)} \\ = (U_{i+1}^j - U_i^j)/h & \text{(3) forward differences, from (6)} \\ = (U_i^j - U_{i-1}^j)/h & \text{(4) backward differences, from (7)} \end{cases}$$

The diagram of the panel shows x and t axes and a network of discrete points formed by orthogonal lines separated by a distance h on the x-axis and k on the t-axis. The function $U = U(x, t)$ will be represented by finite-difference assumptions based on the discrete points formed by the network. The distances h and k are shown as constants although this restriction is not necessary and often undesirable.

The finite difference formulas, obtained by Taylor series expansions, are presented in the panel in notation which will be used subsequently and which readily allows the identification of mesh points. Thus, the value of U at the mesh point (x_i, t_j) is written as

$$U(x_i, t_j) = U_i^j$$

and so on for other points on the mesh.

In the context of the present parabolic equation, with the boundary and initial condition requirements of panel 5.3, a second vertical axis could be chosen at $x = L$ and the solution domain would then be seen to be confined on three sides. This is shown on the following two panels.

5.7 Explicit method

- Solve

$$\frac{\partial U}{\partial t} = \frac{\partial^2 U}{\partial x^2}$$

  □ Unknown
  × Known
  ○ "centring"

- Explicit Method

$$\frac{U_i^{j+1} - U_i^j}{k} = \frac{1}{h^2}[U_{i+1}^j - 2U_i^j + U_{i-1}^j]$$

- Solving for U_i^{j+1} we get

$$U_i^{j+1} = U_i^j + \frac{k}{h^2}(U_{i-1}^j - 2U_i^j + U_{i+1}^j) \quad i = 1, 2, \ldots, I-1$$

- An explicit method is (a) easy to solve, (b) first-order accurate in time and (c) has stability requirements

The solution of partial differential equations by finite difference methods can be obtained by an explicit or an implicit method. The two methods are explained by panels 5.7 and 5.8 which also indicate their relative advantages. The equations and figure of panel 5.7 express one unknown grid value directly in terms of known grid values and correspond to an explicit formulation.

With a forward difference formula, equation (3) of panel 5.6, to approximate $\partial U/\partial t$ and equation (1) of panel 5.6 to approximate $\partial^2 U/\partial x^2$, both at the net point (x_i, t_j), we get the first equation of the above panel. This is an explicit finite difference approximation to the heat conduction equation centred at the net point (x_i, t_j). The error in this approximation is of the order $(k + h^2)$. The value of $U_i^{j+1} = U(x_i, t_{j+1})$ is explicitly given from this approximation by the second equation, and all the new values at $t = t_{j+1}$ can be determined.

Although an explicit method is computationally simple, it has stability requirements; the time step k must be very small, i.e. $k < h^2/2$ and h must also be kept small to maintain reasonable accuracy.

5.8 Crank–Nicholson method (implicit), 1

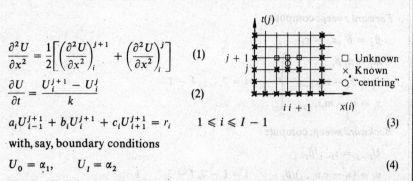

$$\frac{\partial^2 U}{\partial x^2} = \frac{1}{2}\left[\left(\frac{\partial^2 U}{\partial x^2}\right)_i^{j+1} + \left(\frac{\partial^2 U}{\partial x^2}\right)_i^j\right] \quad (1)$$

$$\frac{\partial U}{\partial t} = \frac{U_i^{j+1} - U_i^j}{k} \quad (2)$$

$$a_i U_{i-1}^{j+1} + b_i U_i^{j+1} + c_i U_{i+1}^{j+1} = r_i \quad 1 \leqslant i \leqslant I-1 \quad (3)$$

with, say, boundary conditions

$$U_0 = \alpha_1, \qquad U_I = \alpha_2 \quad (4)$$

Crank–Nicholson method: (a) involves more arithmetic per t step, (b) has second-order accuracy with *uniform* x-spacing and (c) is unconditionally stable

Implicit methods do not require k to be small to achieve stability and this represents a major advantage. As an example of an implicit method, the Crank–Nicholson (1947) method is described in relation to the panel.

The second-order term $\partial^2 U/\partial x^2$ is replaced by the mean of finite difference approximations at $(x_i, t_{j-1/2})$, indicated by 0 in the above panel, and a central difference formulation used for $\partial U/\partial t$ at the same points. These approximations are shown, respectively, in equations (1) and (2) which may be substituted into the steady heat equation to yield

$$\frac{U_i^{j+1} - U_i^j}{k} = \tfrac{1}{2}\left(\frac{U_{i+1}^{j+1} - 2U_i^{j+1} + U_{i-1}^{j+1}}{h^2} + \frac{U_{i+1}^j - 2U_i^j + U_{i-1}^j}{h^2}\right).$$

This equation can also be written in the form of equation (3) of the panel with

$$a_i = -\frac{k}{h^2}, \quad b_i = 2\left(1 + \frac{k}{h^2}\right), \quad c_i = -\frac{k}{h^2},$$

$$r_i = \frac{k}{h^2} U_{i-1}^j + 2\left(1 - \frac{k}{h^2}\right) U_i^j + \frac{k}{h^2} U_{i+1}^j.$$

The error in this scheme is of the order $(k^2 + h^2)$ and k need *not* be related to h for stability, at least for the unsteady heat conduction equation.

5.9 Crank–Nicholson method (implicit), 2

Forward sweep: compute

$\beta_1 = b_1, s_1 = r_1$

$\left.\begin{array}{l} m_i = a_i/\beta_{i-1} \\ \beta_i = b_i - m_i c_{i-1} \\ s_i = r_i - m_i s_{i-1} \end{array}\right\} \quad i = 2, \ldots, I - 1$

Backward sweep: compute

$U_{I-1} = s_{I-1}/\beta_{I-1}$

$u_i = (s_i - c_i u_{i+1})/\beta_i \qquad i = I - 2, I - 3, \ldots, 1$

If there are $I + 1$ grid points along each constant time line or "time row," then equation (3) of panel 5.8 gives $I - 1$ simultaneous equations for the $I - 1$ unknown U values along the time row in terms of known initial and *boundary* values. A method such as this where the calculation of an unknown grid value necessitates the solution of a set of simultaneous equations is called *implicit*.

The solution of the system given by equations (3) and (4) can be obtained by a very efficient procedure (see, for example, Keller, 1968). First, let us write the equations in vector notation as

$$AU = r \qquad (6)$$

where we have introduced the $(I - 1)$ dimensional vectors

$$U \equiv \begin{bmatrix} U_1 \\ U_2 \\ \vdots \\ U_{I-1} \end{bmatrix} \qquad r \equiv \begin{bmatrix} r_1 \\ r_2 \\ \vdots \\ r_{I-1} \end{bmatrix}$$

and the "tridiagonal" matrix of order $I - 1$

$$A = \begin{bmatrix} b_1 & c_1 & 0 & & & & \\ a_2 & b_2 & c_2 & & & & \\ & & \ddots & & & & \\ & & & & a_{I-2} & b_{I-2} & c_{I-2} \\ & & & & & a_{I-1} & b_{I-1} \end{bmatrix}$$

The Solution of the Thin Shear-layer Equations

Then (see, e.g. Keller, 1968) the solution of equation (8) (or (3) and (4)) is obtained by two sweeps as shown in the panel above. The meaning of the word "sweep" here refers to travel along a constant time line, not to a complete calculation for all times as in panel 5.2.

5.10 Box method of Keller (implicit), 1

Features of the Box method:
1. Slightly more complicated arithmetically
2. Second-order accuracy with *arbitrary* (nonuniform) x-spacing
3. Allows very rapid x-variations, even discontinuities

Box Method involves four steps:
1. Reduce the governing equations to a system of first-order ones
2. Write difference equations with central differences
3. Linearize the resulting algebraic equations (if they are non-linear) and write them in matrix-vector form
4. Solve the linear system by the block tridiagonal elimination method

---

An alternative, and more recent, implicit procedure is that described by Keller (1970) and Keller and Cebeci (1972), called the Box method. It has been applied to the solution of thin shear-layer equations with boundary conditions corresponding to a wide range of flow problems and is preferred because it combines convenience of use with numerical precision and economy.

As indicated in the panel, the Box method requires the solution of more algebraic equations than the Crank–Nicholson procedure but maintains second-order accuracy with nonuniform distributions of mesh points. Because the equations are first order, the additional arithmetic is convenient and is easily programmed. It is unconditionally stable and permits the use of large values of h and k. The solution of the tridiagonal system is readily achieved as described briefly in relation to panel 5.13.

The Solution of the Thin Shear-layer Equations

5.11 Box method of Keller (implicit), 2

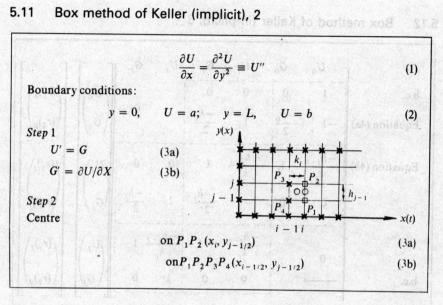

$$\frac{\partial U}{\partial x} = \frac{\partial^2 U}{\partial y^2} \equiv U'' \qquad (1)$$

Boundary conditions:
$$y = 0, \quad U = a; \quad y = L, \quad U = b \qquad (2)$$

Step 1
$$U' = G \qquad (3a)$$
$$G' = \partial U/\partial X \qquad (3b)$$

Step 2
Centre

on $P_1 P_2 \,(x_i, y_{j-1/2})$ \qquad (3a)

on $P_1 P_2 P_3 P_4 \,(x_{i-1/2}, y_{j-1/2})$ \qquad (3b)

Equation (1) is similar to the unsteady heat conduction equation but, with x and y as independent variables, is close to the form which will be solved for thin shear layers.

Step 1 of the panel reduces equation (1) to a system of two first-order equations as shown by equations (3a) and (3b). It requires the introduction of new variables and will be discussed further in relation to panel 6.2.

Following the procedure of step 2, we write the difference equations for the first-order system as shown in the panel. Note here that checks ($\times$) denote the known solution, squares ($\square$) the unknown solution. Also, circles ($\bigcirc$) denote the location of centring of the differential equation. We use the notation that for any quantity g

$$g_j^{i-1/2} = \tfrac{1}{2}(g_j^i + g_j^{i-1}), \qquad g_{j-1/2}^i = \tfrac{1}{2}(g_j^i + g_{j-1}^i), \qquad (g')_{j-1/2}^i = \frac{g_j^i - g_{j-1}^i}{h_{j-1}}.$$

The difference equations for equations (3a) and (3b) are

$$\frac{U_j^i - U_{j-1}^i}{h_{j-1}} = G_{j-1/2}^i \rightarrow U_j^i - U_{j-1}^i - \frac{h_{j-1}}{2}(G_j^i + G_{j-1}^i) = 0, \qquad (4a)$$

$$G_j^i - G_{j-1}^i - \frac{h_{j-1}}{k_i}(U_j^i + U_{j-1}^i) = h_{j-1} R_{j-1/2}^{i-1}, \qquad (4b)$$

where

$$R_{j-1/2}^{i-1} = (G')_{j-1/2}^{i-1} - \frac{2}{k_i} U_{j-1/2}^{i-1}. \qquad (4c)$$

The boundary conditions, equation (2), become:
$$U_0 = a, \qquad U_J = b.$$

5.12 Box method of Keller (implicit), 3

	U_0	G_0	U_j	G_j	U_J	G_J			
b.c.	1	0	0	0		0	$\begin{pmatrix}U_0\\G_0\end{pmatrix}$	=	$\begin{pmatrix}(r_1)_0\\(r_2)_0\end{pmatrix}$
Equation (4a)	-1	$\dfrac{-h_0}{2}$	1	$\dfrac{-h_0}{2}$					
Equation (4b)	$\dfrac{-h_{j-1}}{k_i}$	-1	$\dfrac{-h_{j-1}}{k_i}$	1	0	0	$\begin{pmatrix}U_j\\G_j\end{pmatrix}$		$\begin{pmatrix}(r_1)_j\\(r_2)_j\end{pmatrix}$
	0	0	-1	$\dfrac{-h_j}{2}$	1	$\dfrac{-h_j}{2}$			
	0		$\dfrac{-h_{j-1}}{k_i}$	-1	$\dfrac{-h_{j-1}}{k_i}$	1	$\begin{pmatrix}U_J\\G_J\end{pmatrix}$		$\begin{pmatrix}(r_1)_J\\(r_2)_J\end{pmatrix}$
b.c.	$\longrightarrow$		0	0	1	0			

Since the system under study is a linear one, the linearization procedure of step 3 is not required and equation (4) can be written in matrix-vector form as shown in the panel. Here the first and last rows correspond to the boundary conditions, which are reflected in the definitions of $(r_1)_0$ and $(r_2)_J$. The right-hand sides of equations (4), including the boundary conditions, are given by

$$(r_1)_0 = a, \qquad (r_1)_j = h_{j-1} R_{j-1/2}^{i-1} \quad (1 \leqslant j \leqslant J),$$
$$(r_2)_j = 0 \quad (0 \leqslant j \leqslant J-1), \qquad (r_2)_J = b. \tag{5}$$

Note that for compactness we have shown only the values of the dependent variables at boundaries corresponding to $j = 0, J$. Intermediate values are, of course, represented by j.

5.13 Box method of Keller (implicit), 4

Here
$$A\delta = r \qquad (6)$$

$$A \equiv \begin{bmatrix} A_0 & C_0 & & & & \\ B_1 & A_1 & C_1 & & & \\ & B_j & A_j & C_j & & \\ & & B_{J-1} & A_{J-1} & C_{J-1} & \\ & & & & B_J & A_J \end{bmatrix}, \quad \delta = \begin{bmatrix} \delta_0 \\ \delta_1 \\ \vdots \\ \delta_j \\ \vdots \\ \delta_J \end{bmatrix}, \quad r = \begin{bmatrix} r_0 \\ r_1 \\ \vdots \\ r_j \\ \vdots \\ r_J \end{bmatrix}$$

$$\delta_j = \begin{bmatrix} U_j \\ G_j \end{bmatrix}, \quad r_j = \begin{bmatrix} (r_1)_j \\ (r_2)_j \end{bmatrix} \text{ and } A_j, B_j, C_j \text{ are } 2 \times 2 \text{ matrices}$$

As in panel 5.9, the system (4) of panel 5.11 can also be written in compact form as shown by (6) in the above panel.

The definitions of 2×2 matrices A_j, B_j, C_j are

$$A_0 \equiv \begin{pmatrix} 1 & 0 \\ -1 & \frac{-h_0}{2} \end{pmatrix} \quad C_j \equiv \begin{pmatrix} 0 & 0 \\ 1 & \frac{-h_j}{2} \end{pmatrix} \quad 0 \leq j \leq J-1 \quad A_J \equiv \begin{pmatrix} \frac{-h_J}{k_i} & 1 \\ 1 & 0 \end{pmatrix},$$

$$B_j \equiv \begin{pmatrix} \frac{-h_{j-1}}{k_i} & -1 \\ 0 & 0 \end{pmatrix} \quad 1 \leq j \leq J \quad A_j \equiv \begin{pmatrix} \frac{-h_{j-1}}{k_i} & 1 \\ -1 & \frac{-h_j}{2} \end{pmatrix} \quad 1 \leq j \leq J-1.$$

Again, as in panel 5.9, the implicit nature of the method has generated a tridiagonal matrix, but with 2×2 blocks rather than scalars.

The solution of equation (6) is obtained by the block elimination method discussed by Keller (1974) and by Cebeci and Bradshaw (1977). This method consists of two sweeps. In the *forward sweep*, Γ_j, Δ_j and w_j are computed from the following recursion formulas:

$$\begin{aligned} \Delta_0 &= A_0 \\ \Gamma_j \Delta_{j-1} &= B_j \\ \Delta_j &= A_j - \Gamma_j C_{j-1} \end{aligned} \biggr\} 1 \leq j \leq J$$

$$\begin{aligned} w_0 &= r_0 \\ w_j &= r_j - \Gamma_j w_{j-1} \quad 1 \leq j \leq J. \end{aligned}$$

In the backward sweep δ_j is obtained from the recursion formulas:

$$\Delta_J \delta_J = w_J$$
$$\Delta_j \delta_j = w_j - C_j \delta_{j+1} \quad j = J-1, J-2, \ldots, 0,$$

5.14 Thin shear-layer equations for two-dimensional flows and axisymmetric flows

$$-\overline{\rho uv} = \rho v_t \frac{\partial U}{\partial y} \quad (1)$$

$$\frac{\partial}{\partial x}(rU) + \frac{\partial}{\partial y}(rV) = 0 \quad (2)$$

$$U\frac{\partial U}{\partial x} + V\frac{\partial U}{\partial y} = -\frac{1}{\rho}\frac{dP}{dx} + \frac{1}{r}\frac{\partial}{\partial y}\left[r(v + v_t)\frac{\partial U}{\partial y}\right] \quad (3)$$

Here

$$-\frac{1}{\rho}\frac{dP}{dx} = U_e \frac{dU_e}{dx}$$

The following two chapters, and part of Chapter 9, are concerned with the two-dimensional, steady form of the thin shear-layer equations for laminar and turbulent flows. Equations (2) and (3) of the panel represent conservation of mass and momentum. For a plane two-dimensional flow $r = 1$ and for an axisymmetric flow $r = r_0 + y \cos \alpha$, where $r_0(x)$ is the body radius and $\alpha = \tan^{-1} dr_0/dz$, z being the coordinate along the axis of symmetry. In dimensionless form r can be written as

$$\frac{r}{r_0} = 1 + \frac{y \cos \alpha}{r_0} = 1 + t.$$

Here t represents the deviation of r from r_0 and is called the transverse curvature term. In many axisymmetric flows the body radius is quite large in relation to the boundary-layer thickness, so the transverse curvature effect is negligible. More will be said in relation to panel 6.6.

Equation (3) presumes that an eddy viscosity assumption, equation (1), will be used and when evaluated for large y (or r) leads to the last equation of the panel, a one-dimensional form of Bernoulli's equation.

5.15 Boundary conditions for two-dimensional, thin shear-layer equations, 1

---

a. External flows (no mass transfer)

$y = 0, \quad U = 0, \quad V = 0,$ (1a)

$y = \delta, \quad U = U_e(x)$ (1b)

Other "wall" boundary conditions (at $y = y_0$)

$$U_0 = U_\tau \left[\frac{1}{\kappa} \ln \frac{y_0 U_\tau}{\nu} + 5 \cdot 2 \right]$$ (2a)

$$V_0 = -\frac{U_0 y_0}{U_\tau} \frac{dU_\tau}{dx}, \quad U_\tau = \sqrt{\frac{\tau_w}{\rho}}$$ (2b)

$$\frac{\tau_0}{\rho} = \frac{\tau_w}{\rho} + \frac{1}{\rho} \frac{dP}{dx} y_0 + \frac{\alpha^*}{\rho} \frac{d\tau_w}{dx} y_0$$ (3)

The usual boundary conditions for the thin shear-layer equations (for both laminar and turbulent external flows) are given by equation (1) of the panel. In turbulent flows, to save computer storage, the inner boundary conditions, equation (1a), may be applied at the first mesh point from the surface (outside the viscous sublayer, say at $y^+ (\equiv y_0 U_\tau/\nu)$ of 50). There are several versions of these "inner" boundary conditions and in the above panel, we show the version of Cebeci, Chang and Bradshaw (1980). Note that, when equations (2a) and (2b) are used as a boundary condition, at first U_τ is not known, and must be guessed. Once a solution is obtained, however, τ_0/ρ is known and equation (3) can be used to compute a new value of τ_w/ρ. An alternative procedure is to use the Mechul method to be discussed in Chapter 7.

In equation (3), α^* depends on the velocity profile for $y < y_0$ and is given by

$$\alpha^* = \tfrac{1}{2}[K_1 (\ln y_0^+)^2 + K_2 \ln y_0^+ + K_3 + K_4/y_0^+]$$ (4)

where $K_1 = 5 \cdot 95$, $K_2 = 13 \cdot 5$, $K_3 = 15 \cdot 6$ and $K_4 = -698$. An extension of equation (3) to two-dimensional unsteady flows is described in Cebeci, Carr and Bradshaw (1979).

5.16 Boundary conditions for two-dimensional, thin shear-layer equations, 2

b. Internal flows (i.e. parallel-plate channel)

$y = 0: U = 0, V = 0$

$y = h: \dfrac{\partial U}{\partial y} = 0$

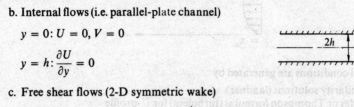

c. Free shear flows (2-D symmetric wake)

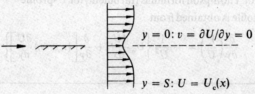

$y = 0: v = \partial U/\partial y = 0$

$y = S: U = U_c(x)$

In internal flows $P(x)$ is not known prior to the calculations and must be determined during the calculations in such a way that the solution generated for a given $P(x)$ will satisfy conservation of mass. If we consider the flow between two parallel plates, then constant mass flow rate means constant U_0, where

$$U_0 h = \int_0^h U \, dy.$$

As we shall see later in Chapter 7, for a developing flow it may be more convenient to use slightly different edge boundary conditions than those shown above.

In the calculation of the wake downstream of a wall boundary-layer flow, numerical difficulties can arise due to the jump in the inner or wall boundary conditions since, for example, U is zero on the body and is nonzero on the centreline of the wake. For either a laminar or turbulent flow, the change in boundary conditions generates an inner layer near the centreline of the wake; this complication can be treated numerically with a form of double structure, as described by Cebeci, Thiele, Williams and Stewartson (1979).

5.17 Initial conditions for two-dimensional steady external flows

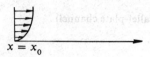
$x = x_0$

- Initial conditions are generated by
  a. similarity solutions (laminar)
  b. Coles or Thompson formulas (turbulent) for U-profile
- The V-profile is obtained from

$$\frac{\partial}{\partial y}\left(\frac{V}{U}\right) = -\frac{1}{U^2}\left\{U_e\frac{dU_e}{dx} + \frac{\partial}{\partial y}\left[(v+v_t)\frac{\partial U}{\partial y}\right]\right\} \quad (1)$$

For two-dimensional wall boundary-layer flows, initial conditions may be determined by specifying initial velocity profiles. Often similarity solutions are used to generate the initial conditions for laminar flows and several transformations can be used for this purpose, including that of Falkner and Skan.

A popular formula used to generate the initial U-velocity profile for turbulent flows is that given by Coles (1956). The normal component of the velocity, V, can be obtained from the x-equation (1) which results from momentum equation and the continuity equation.

Generation of initial turbulent velocity profiles from Coles formula for three-dimensional flows does not lead to a good fit with experimental data. For this reason, it is more appropriate to use Thompson's profiles (see Galbraith and Head, 1975) for three-dimensional and even for two-dimensional flows. According to this formula the dimensionless velocity profile is given by

$$\frac{U}{U_e} = \gamma_s\left(\frac{U}{U_e}\right)_{inner} + (1 - \gamma_s), \quad (2)$$

where γ_s is an intermittency factor defined by the empirical formulas:

$$0 < \frac{y}{\delta_0} \leq 0.05, \quad \gamma_s = 1,$$

$$0.05 < \frac{y}{\delta_0} \leq 0.3, \quad \gamma_s = 1 - 2.64\left(\frac{y}{\delta_0} - 0.05\right)^2,$$

$$0.3 < \frac{y}{\delta_0} \leq 0.7, \quad \gamma_s = 4.40\left(\frac{y}{\delta_0} - 0.5\right)^3 - 1.85\left(\frac{y}{\delta_0} - 0.5\right) + 0.5,$$

The Solution of the Thin Shear-layer Equations

$$0.7 < \frac{y}{\delta_0} \leq 0.95, \quad \gamma_s = 2.64\left(\frac{y}{\delta_0} - 0.05\right)^2,$$

$$\frac{y}{\delta_0} > 0.95, \quad \gamma_s = 0.0. \tag{3}$$

The dimensionless velocity profile for the inner layer, that is, $U^+(y^+)$, is given by

$$y^+ < 4, \quad U^+ = y^+,$$
$$4 < y^+ < 30, \quad U^+ = c_1 + c_2 \ln y^+ + c_3(\ln y^+)^2 + c_4(\ln y^+)^3,$$
$$y^+ > 30, \quad U^+ = 5.50 \log_{10} y^+ + 5.45. \tag{4}$$

Here $c_1 = 4.19$, $c_2 = -5.75$, $c_3 = 5.11$, $c_4 = -0.767$, $y^+ = yU_\tau/\nu$, $U_\tau = (\tau_w/\rho)^{1/2}$, $U^+ = U/U_\tau$ and δ_0 is a parameter which is a function of θ, c_f and H.

To find the functional relationship between δ_0, c_f, θ and H, we use the definitions of displacement thickness δ^* and momentum thickness θ. Substituting (2) into the definition of δ^* gives, after some algebra,

$$\frac{\delta^*}{\delta_0}\left(1 - \frac{A_1}{R_{\delta^*}}\right) = 0.5 + \sqrt{\frac{c_f}{2}}\left[A_4 \ln \frac{\delta^*}{\delta_0} - A_3 - A_2 \ln\left(R_{\delta^*}\sqrt{\frac{c_f}{2}}\right)\right] \tag{5}$$

where

$$A_1 = 50.68, \quad A_2 = 1.194, \quad A_3 = 0.794, \quad A_4 = 1.195.$$

An expression similar to that given by (5) can also be obtained by substituting (2) into the definition of θ. However, the resulting expression is complicated, so we obtain the expression for θ numerically and, for given values of θ and H, compute the corresponding values of c_f and δ_0 from equations (2) and (5).

Equation (2) is recommended for two-dimensional flows. It can also be assumed to apply to the streamwise velocity profile of a three-dimensional flow by replacing U/U_e with U_s/U_{se} and letting c_f represent the streamwise skin-friction coefficient. In order to generate the crossflow velocity component (U_n/U_{se}), we use Mager's expression and define U_n/U_{se} by

$$\frac{U_n}{U_{se}} = \frac{U_s}{U_{se}}\left(1 - \frac{y}{\delta}\right)^2 \tan \beta_w \tag{6}$$

with the limiting crossflow angle β_w obtained from the experimental data.

5.18 Initial conditions for two-dimensional unsteady and three-dimensional steady flows

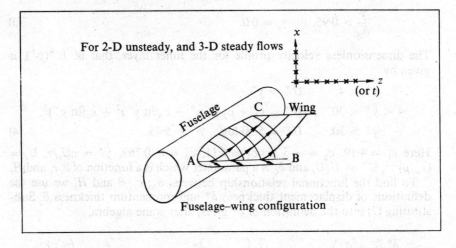

Fuselage–wing configuration

For three-dimensional steady flows, and two-dimensional unsteady flows, it is necessary to generate initial conditions on two intersecting lines as indicated in the panel. In some problems they can be established with ease; in others they require careful and extensive studies. As an example, consider the fuselage–wing combination shown in the above panel. Here the spanwise initial conditions (AB) are generated by solving a set of special equations, called the stagnation-line equations, constructed by taking into account the symmetry conditions. However, the initial conditions on AC, which forms the wing-fuselage junction, cannot be calculated easily. In fact, the viscous flow along the line AC is not of the boundary-layer type and corner-flow equations must be solved (see Chapter 11). The problem is complicated by the presence of secondary flows in the plane orthogonal to AC, which stem mainly from the Coriolis forces due to the curvature of the wing but may also be influenced by stress-driven secondary flows. A satisfactory calculation method does not exist for the secondary flow although related work, such as that described in Chapters 10 and 11, is in progress, and the boundary condition on AC has therefore to be estimated and the influence of different estimates obtained by calculation. This topic will be discussed further in Chapter 13 in connection with wing flows.

5.19 Definition of standard and inverse boundary-layer problems

---

Standard boundary-layer problem

Solve thin shear-layer equations subject to, say

$$y = 0, \quad U = 0, \quad V = V_w(x)$$
$$y = \delta, \quad U = U_e(x) \quad \text{(given)}$$

Inverse boundary-layer problem

Solve thin shear-layer equations subject to, say

$$y = 0 \quad U = 0 \quad V = V_w(x)$$
$$y = \delta \quad U = U_e(x) \quad \text{(unknown)}$$

$$\delta^*(x) = \int_0^\delta \left(1 - \frac{U}{U_e}\right) dy = \text{(given)}$$

Boundary-layer calculations are usually performed for prescribed boundary conditions which, for two-dimensional incompressible flows may be provided in the form

$$U(x, 0) = 0, \quad V(x, 0) = V_w(x), \quad U(x, \delta) = U_e(x).$$

For convenience we shall call them *standard* boundary-layer problems. In some problems, however, the external velocity distribution, or for internal flows the pressure drop, is unknown and can be determined to satisfy an alternative boundary condition, such as prescribed displacement thickness $\delta^*(x)$, prescribed wall shear $\tau_w(x)$ or, for internal flows, conservation of mass. The solutions to these problems can be obtained by "inverse boundary-layer methods" and find application in many practical problems. For example, the airfoils discussed by Liebeck (1978) are designed on the principle that, in certain regions of an airfoil flow, the deceleration produces values of wall-shear stress close to zero. Other important applications of inverse boundary-layer procedures occur in the calculation of duct flows such as those discussed in Chapter 3. They also occur in the calculation of separated flows such as those discussed by Carter (1975), Cebeci (1976a, b), Williams (1975), Cebeci, Keller and Williams (1979), Cebeci, Khalil and Whitelaw (1979).

5.20 Concluding remarks

This brief introduction to finite-difference methods has been related particularly to parabolic equations and to boundary and initial conditions appropriate to boundary-layer flows. Elliptic equations, such as the steady form of the Navier–Stokes equations, are considered in Chapter 10.

The solution of external two- and three-dimensional wall boundary-layer flows has become almost a matter of routine in some industries and research groups and is comparatively straightforward. The following chapter indicates the requirements in some detail and gives a glimpse of what has been achieved. Inverse problems, and especially those involving flow separation, are far from routine, and Chapter 7 is concerned with them and the ways in which they can be solved. Similarly, unsteady two-dimensional and steady three-dimensional flows, particularly those with negative cross-flow velocities, require special treatment, and Chapter 8 is devoted to them.

6 Solution of Two-dimensional, External Boundary-layer Problems

6.1 Purposes and outline of the chapter

- To present Box method of Keller for steady two-dimensional flow
- To discuss Mangler and Falkner–Skan transformations
- To emphasize the importance and prediction of transition
- To present a double-structured numerical method appropriate for solving parabolic equations with discontinuous boundary conditions

This chapter is concerned with the application of the Box method to the solution of boundary-layer problems. It is limited to steady, external, two-dimensional boundary layers which may be plane or axisymmetric. The following three panels indicate and discuss the transformation of the thin shear-layer equations to a first-order set and their representation, in finite-difference form, by the Box method. Panel 6.5 is concerned with the treatment of boundary conditions within this framework.

The Mangler and Falkner–Skan transformations can each make useful contributions to the solution of boundary-layer problems. The former allows axisymmetric and plane, two-dimensional problems to be represented by the same equations and the latter can assist with the generation of initial conditions, with the removal of the singularity at $x = 0$ and in maintaining a boundary layer of near constant thickness for most laminar flows. These properties are not essential but can be useful. For example, a coordinate system which leads to a boundary layer of constant thickness helps to permit the use of a numerical method with constant precision and economy.

As a typical "standard" boundary-layer problem, the flow on an airfoil is considered; this allows us to discuss practical methods for the prediction of transition as well as the calculation of total drag and the downstream wake. The wake introduces a singularity which requires special attention; one method of solution, involving a double-structured arrangement, is described in panels 6.12 to 6.15.

6.2 Box method for the standard problem, 1

> The thin shear-layer equations with boundary conditions (no mass transfer)
> $$U = V = 0 \quad \text{at } y = 0; \quad U = U_e(x) \quad \text{at } y = \delta \qquad (1)$$
> can be written as the first-order system:
> a. $\psi' = U$
> b. $U' = G$ (2)
> c. $(bG)' - \dfrac{d\overline{P}}{dx} = U \dfrac{\partial U}{\partial x} - G \dfrac{\partial \psi}{\partial x}$
> d. $y = 0, \quad \psi = U = 0, \quad y = \delta, \quad U = U_e$
> where ψ is the stream function, $\overline{P} = P/\rho$ and $b = v + v_t$

As indicated in the previous chapter, Keller's Box method has proved to be particularly convenient, economic and accurate and is selected here for special attention. The main steps in its use are documented on this and the following pages.

For two-dimensional incompressible plane flows, the thin shear-layer equations, with the eddy viscosity concept, may be written in the form

$$\frac{\partial U}{\partial x} + \frac{\partial V}{\partial y} = 0,$$

$$U \frac{\partial U}{\partial x} + V \frac{\partial U}{\partial y} = -\frac{1}{\rho} \frac{dP}{dx} + \frac{\partial}{\partial y}\left(b \frac{\partial U}{\partial y}\right).$$

With the introduction of the usual definition of stream function, giving

$$U = \frac{\partial \psi}{\partial y}, \quad V = -\frac{\partial \psi}{\partial x},$$

the continuity and momentum equations can be written as

$$\frac{\partial \psi}{\partial y} \frac{\partial^2 \psi}{\partial x \partial y} - \frac{\partial \psi}{\partial x} \frac{\partial^2 \psi}{\partial y^2} = -\frac{1}{\rho} \frac{dP}{dx} + \frac{\partial}{\partial y}\left(b \frac{\partial^2 \psi}{\partial y^2}\right).$$

This equation is third order and may be written, with primes denoting differentiation with respect to y, as the first-order system shown in the panel with its boundary conditions.

External Boundary-layer Problems

6.3 Box method for the standard problem, 2

- Equations (2a, b) from previous panel are centred at $(x_i, y_{j-\frac{1}{2}})$ on P_1P_2
- Equation (2c) from previous panel is centred at $(x_{i-\frac{1}{2}}, y_{j-\frac{1}{2}})$ on $P_1P_2P_3P_4$
- Difference equations for equations (2a, b, c) are

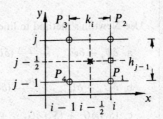

a. $\dfrac{\psi_j^i - \psi_{j-1}^i}{h_{j-1}} = U_{j-\frac{1}{2}}^i$ \quad b. $\dfrac{U_j^i - U_{j-1}^i}{h_{j-1}} = G_{j-\frac{1}{2}}^i$

c. $\dfrac{(bG)_j^i - (bG)_{j-1}^i}{h_{j-1}} + \alpha_i[(\psi G)_{j-\frac{1}{2}}^i + G_{j-\frac{1}{2}}^{i-1}\psi_{j-\frac{1}{2}}^i - \psi_{j-\frac{1}{2}}^{i-1}G_{j-\frac{1}{2}}^i]$

$- (U^2)_{j-\frac{1}{2}}^i] = R_{j-\frac{1}{2}}^{i-1}$ \hfill (3)

Following the procedure of step 2 of panel 5.11, the first-order equations are centred and rewritten in finite-difference form. A typical finite-difference mesh is shown on the panel. The ordinary differential equations (2a, b) of panel 6.2 are centred between the grid points P_1P_2 and the partial differential equation (2c) is centred at the centre of the mesh denoted by $P_1P_2P_3P_4$.

For convenience, $R_{j-\frac{1}{2}}^{i-1}$ has been introduced to represent

$$\alpha_i[(\psi G)_{j-\frac{1}{2}}^{i-1} - (U^2)_{j-\frac{1}{2}}^{i-1} - (U_e^2)^i + (U_e^2)^{i-1}],$$

where

$$\alpha_i \equiv \dfrac{1}{x_i - x_{i-1}}.$$

6.4 Box method for the standard problem, 3

Use Newton's method to linearize equations (3) of panel 6.3 for $1 \leqslant j \leqslant J$:

a. $\delta\psi_j - \delta\psi_{j-1} - \dfrac{h_{j-1}}{2}(\delta U_j + \delta U_{j-1}) = (r_1)_j$

b. $\delta U_j - \delta U_{j-1} - \dfrac{h_{j-1}}{2}(\delta G_j + \delta G_{j-1}) = (r_3)_j$ (4)

c. $(S_1)_j \delta G_j + (S_2)_j \delta G_{j-1} + (S_3)_j \delta\psi_j + (S_4)_j \delta\psi_{j-1} + (S_5)_j \delta U_j + (S_6)_j \delta U_{j-1}$
$= (r_2)_j$

Boundary conditions become ($j = 0$ and $j = J$)

$\delta\psi_0 = (r_1)_1 \equiv 0, \qquad \delta U_0 = (r_2)_1 \equiv 0, \qquad \delta U_J = (r_3)_J \equiv 0$ (5)

The finite difference equations of panel 6.3, namely equations (3), are non-linear so, following the procedure of step 3 of panel 5.11, we linearize them as well as their boundary conditions. We use Newton's method for this purpose as described, for example, by Cebeci and Bradshaw (1977) and introduce the iterates $[\psi_j^{(n)}, U_j^{(n)}, G_j^{(n)}]$, $n = 0, 1, 2, \ldots$, with initial values, say,

$$\psi_0^{(0)} = 0, \qquad U_0^{(0)} = 0, \qquad G_0^{(0)} = G_0^{i-1},$$
$$\psi_j^{(0)} = \psi_j^{i-1}, \qquad U_j^{(0)} = U_j^{i-1}, \qquad G_j^{(0)} = G_j^{i-1}, \qquad 1 \leqslant j \leqslant J-1,$$
$$\psi_J^{(0)} = \psi_J^{i-1}, \qquad U_J^{(0)} = U_e, \qquad G_J^{(0)} = G_J^{i-1}.$$

For the higher iterates we set

$$\psi_j^{(n+1)} = \psi_j^{(n)} + \delta\psi_j^{(n)}, \qquad U_j^{(n+1)} = U_j^{(n)} + \delta U_j^{(n)}, \qquad G_j^{(n+1)} = G_j^{(n)} + \delta G_j^{(n)}.$$

Then we insert the right-hand sides of these expressions in place of ψ_j, U_j, and G_j in equations (3) of panel 6.3 and drop the terms that are quadratic in $(\delta\psi_j^{(n)}, \delta U_j^{(n)}, \delta G_j^{(n)})$. This procedure yields the *linear* system shown on the panel (the superscript n in δ quantities is dropped for simplicity). Here the $(r_k)_j$ ($k = 1, 2, 3$) are defined by

$$(r_1)_j = \psi_{j-1}^i - \psi_j^i + h_{j-1} U_{j-\frac{1}{2}}^i, \qquad (r_3)_j = U_{j-1}^i - U_j^i + h_{j-1} G_{j-\frac{1}{2}}^i,$$
$$(r_2)_j = R_{j-\frac{1}{2}}^{i-1} - \{(bG)'_{j-\frac{1}{2}} + \alpha_i[(\psi G)_{j-\frac{1}{2}} + G_{j-\frac{1}{2}}^{i-1}\psi_{j-\frac{1}{2}} - \psi_{j-\frac{1}{2}}^{i-1} G_{j-\frac{1}{2}} - (U^2)_{j-\frac{1}{2}}]\}^i,$$

and the $(S_k)_j$ ($k = 1$ to 6) by

$$(S_1)_j = \dfrac{b_j}{h_{j-1}} + \dfrac{\alpha_i}{2}(\psi_j^i - \psi_{j-\frac{1}{2}}^{i-1}), \qquad (S_2)_j = -\dfrac{b_{j-1}}{h_{j-1}} + \dfrac{\alpha_i}{2}(\psi_{j-1}^i - \psi_{j-\frac{1}{2}}^{i-1}),$$

$$(S_3)_j = \frac{\alpha_i}{2}(G_j^i + G_{j-\frac{1}{2}}^{i-1}), \qquad (S_4)_j = \frac{\alpha_i}{2}(G_{j-1}^i + G_{j-\frac{1}{2}}^{i-1}),$$

$$(S_5)_j = -\alpha_i U_j, \qquad (S_6)_j = -\alpha_i U_{j-1}.$$

We next write the resulting linear system in matrix–vector form

$$A\boldsymbol{\delta}_j = \mathbf{r}_j$$

and solve it by the block-elimination method discussed in panel 5.13. Care should be taken in ordering the equations in matrix–vector form so that the A_0 matrix is not singular.

Note that the first two rows of the A_0 matrix and the last row of the A_J matrix correspond to the boundary conditions.

6.5 Treatment of nonlinear boundary conditions

$$g_1(\psi_0, U_0, G_0) = 0 \qquad (1)$$
$$g_2(\psi_0, U_0, G_0) = 0 \qquad (2)$$

Linearization of equations (1) and (2) yields

$$\lambda_1 \delta\psi_0 + \lambda_3 \delta U_0 + \lambda_5 \delta G_0 = (r_1)_0 \qquad (3)$$
$$(S_1)_0 \delta\psi_0 + (S_3)_0 \delta U_0 + (S_5)_0 \delta G_0 = (r_2)_0 \qquad (4)$$

The first two rows of the A_0 matrix now become

$$A_0 = \begin{pmatrix} \lambda_1 & \lambda_3 & \lambda_5 \\ (S_1)_0 & (S_3)_0 & (S_5)_0 \\ -1 & -h_0/2 & 0 \end{pmatrix}$$

Recall that the first two rows of the A_0 matrix and the last row of the A_J matrix correspond to boundary conditions. Sometimes, as discussed in panel 5.14, we need to apply them away from the wall, say at $y_0 = 50\nu/U_\tau$. These new boundary conditions can easily be incorporated into the solution procedure discussed in the last three panels. These conditions, namely

$$U_0 = U_\tau \left[\frac{1}{\kappa} \ln \frac{y_0 U_\tau}{\nu} + 5.2 \right], \qquad V_0 = -\frac{U_0 y_0}{U_\tau} \frac{dU_\tau}{dx},$$

together with the equation that gives the relation between the wall shear and the shear stress at y_0 (see panel 5.15), namely

$$\frac{\tau_0}{\rho} = \frac{\tau_w}{\rho} + \frac{1}{\rho}\frac{dP}{dx} y_0 + \alpha^* \frac{1}{\rho}\frac{d\tau_w}{dx} y_0 = (\kappa y_0)^2 \left(\frac{\partial U}{\partial y}\right)_0^2,$$

can be written in the form of equations (1) and (2) in the above panel.

Then after linearizing equations (1) and (2) and writing them in the form given by equations (3) and (4) in the above panel, we can incorporate the two "wall" boundary conditions into the A_0 matrix as shown above. Here the definitions of λ_k, $(S_k)_0$ and $(r_1)_0$ and $(r_2)_0$ are as follows:

$$e_1 = 1 + \frac{2(\alpha^*)^{i-\frac{1}{2}} y_0}{x_i - x_{i-1}},$$

$$e_2 = \frac{\kappa^2 y_0^2 G_0^i}{U_\tau^i e_1},$$

External Boundary-layer Problems

$$e_3 = \frac{(\kappa^2 y_0^2 G_0^2)^i + (\kappa^2 y_0^2 G_0^2)^{i-1} - 2(U_\tau^2)^{i-\frac{1}{2}} - [2y_0/(x_i - x_{i-1})](\bar{P}_i - \bar{P}_{i-1})}{2U_\tau^i e_1}$$

$$- \frac{[2(\alpha^*)^{i-\frac{1}{2}} y_0/(x_i - x_{i-1})]\{(U_\tau^2)^i - (U_\tau^2)^{i-1}\}}{2U_\tau^i e_1},$$

$$\lambda_1 = 0, \quad \lambda_3 = 1\cdot 0,$$

$$\lambda_5 = \left[\frac{1}{\kappa}(\ln y_0^+ + 1) + 5.2\right] e_2,$$

$$(S_1)_0 = 1\cdot 0,$$

$$(S_3)_0 = -\frac{y_0}{2} \frac{U_\tau^1 - U_\tau^{i-1}}{U_\tau^{i-\frac{1}{2}}},$$

$$(S_5)_0 = -\frac{y_0}{2} \frac{U_\tau^{i-1} U_0^{i-\frac{1}{2}}}{(U_\tau^{i-\frac{1}{2}})^2} e_2,$$

$$(r_1)_0 = U_\tau^i\left(\frac{1}{\kappa}\ln y_0^+ + C\right) - U_0 + e_3\left[\frac{1}{\kappa}(\ln y_0^+ + 1) + 5.2\right],$$

$$(r_2)_0 = y_0 \frac{U_0^{i-\frac{1}{2}}}{U_\tau^{i-\frac{1}{2}}}(U_\tau^i - U_\tau^{i-1}) - (\psi_0^i - \psi_0^{i-1}) + \frac{y_0}{2} \frac{U_0^{i-\frac{1}{2}} U_\tau^{i-1}}{(U_\tau^{i-\frac{1}{2}})^2} e_3.$$

Note that in equations (1) and (2) we have used the new dependent variables $\psi' = U, \psi'' = G$, etc.

6.6 Mangler and Falkner–Skan transformations

With the Mangler transformation ($k = 1$) for axisymmetric flow, $k = 0$ for two-dimensional flow)

$$d\bar{x} = \left(\frac{r_0}{L}\right)^{2k} dx, \qquad d\bar{y} = \left(\frac{r}{L}\right)^k dy \tag{1}$$

and the Falkner–Skan transformation

$$\eta = \sqrt{\frac{U_e}{\nu \bar{x}}}\,\bar{y}, \qquad \psi = (\bar{U}_e \nu \bar{x})^{1/2} f(\bar{x}, \eta) \tag{2}$$

the continuity and momentum equations of panel 5.14 become

$$(bf'')' + \frac{m+1}{2} ff'' + m[1 - (f')^2] = \bar{x}\left(f'\frac{\partial f'}{\partial \bar{x}} - f''\frac{\partial f}{\partial \bar{x}}\right)$$

$$\eta = 0, \quad f' = 0, \quad f = f_w \tag{4}$$
$$\eta = \eta_e, \quad f' = 1$$

$$m = \bar{x}/\bar{U}_e\, d\bar{U}_e/d\bar{x} \qquad b = (1 + \nu_t/\nu) \tag{5}$$

On the next panel we begin consideration of the particular problem of the calculation of flow on an airfoil. The thin shear-layer equations are written in rectangular Cartesian coordinates and subsequently transformed, as indicated below, to try to maintain a nearly constant boundary-layer thickness for laminar flows and to provide initial conditions.

Similar problems on axisymmetric bodies including the transverse curvature effect if needed, can be represented by the same equations if the Mangler transformation of the panel is used.

As is well known, the Falkner–Skan transformation is appropriate to laminar wedge flows, for which it converts the partial differential thin shear-layer equations into ordinary differential equations. For laminar boundary layers which do not yield similar solutions, and for turbulent boundary layers, the transformation is still useful since it represents the growth of a similar laminar boundary layer (which may not be too far from that of a non-similar layer) and has the effect of reducing the growth of the boundary layer, in the transformed coordinates, so that the number and distribution of finite-difference grid points selected for the initial profile will still be appropriate to a profile at a far-downstream station.

The solution procedure utilizing the Box method can easily be used to solve the system (3) and (4) by the procedure described in panels 6.2, 6.3 and 6.4. For example, (3a, b) of panel 6.3 and (4a, b) of panel 6.4 remain the same. Only 3(c) of panel 6.3 and 4(c) of panel 6.4 are different. However, equation (3) of the above panel can easily be put into the form given by equation 3(c) of panel 6.4. Thus, to solve the system (3) and (4), all we need to do is to define new $S_k (k = 1$ to 6) and $(r_2)_j$. The rest of the solution procedure remains essentially unchanged.

6.7 A typical standard boundary-layer problem: flow along an airfoil given the pressure distribution

Boundary-layer flow along an airfoil

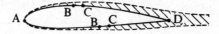

For two-dimensional incompressible flow, a convenient v_t formulation is

$$v_t = \begin{cases} (v_t)_i = l^2 \left|\dfrac{\partial U}{\partial y}\right| \gamma_{tr} \\ (v_t)_o = \alpha \left|\displaystyle\int_0^\infty (U_e - U)\,dy\right| \gamma_{tr} \end{cases} \qquad \dfrac{v_t}{U_e \delta^*}\underline{\qquad\qquad} \quad (1)$$

$$y/\delta$$

where

$l = 0{\cdot}4\, y[1 - \exp(-y/A)]$

Usually, $\alpha = 0{\cdot}0168$, $A = 26\, v/U_\tau$

The flow past an airfoil has practical relevance and is used here as a specific example of a standard boundary-layer problem. It can be divided into four regions corresponding to the laminar flow from A to B, transition from B to C, the turbulent wall flow from C to D and the downstream wake. For the purpose of this discussion, the angle of attack is sufficiently small so there is no separation near the trailing edge of the suction surface.

The calculation starts at A. The system of equations (3) and (4) from panel 6.6 (with flow index $k = 0$) are solved numerically with $v_t = 0$ until the value of x corresponds to transition, which can either be input or computed by the empirical methods to be discussed in panel 6.9. Alternatively, the Cebeci–Smith eddy viscosity formulation of panel 3.6 can be used, allowing the computation of the laminar, transition and turbulent wall-flow regions without a further correlation equation. The expression used for the transitional region in the eddy viscosity formulation is discussed further on the following panel.

External Boundary-layer Problems

6.8 Transition region formulation

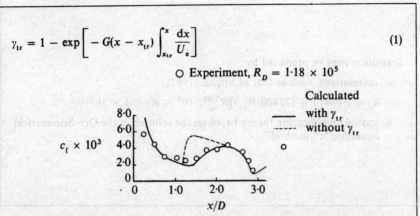

$$\gamma_{tr} = 1 - \exp\left[-G(x - x_{tr})\int_{x_{tr}}^{x}\frac{dx}{U_e}\right] \qquad (1)$$

○ Experiment, $R_D = 1.18 \times 10^5$

Calculated
—— with γ_{tr}
----- without γ_{tr}

Comparison of calculated and experimental skin-friction coefficients for Schubauer's (1939) ellipse

The parameter γ_{tr} in equation (1) is an intermittency factor that accounts for the transitional region between a laminar and a turbulent flow. For two-dimensional flows it is given by equation (1) in the above panels. It can also be used for axisymmetric flows with a minor modification, as shown by Cebeci and Smith (1974).

Inclusion of this expression in the eddy viscosity formulas is especially important at low Reynolds numbers (see equation (2) and also panel 3.6). Here $R_{\Delta x}$ is based on the length of the transition region and G is a spot-formation-rate parameter defined by

$$G = \frac{1}{1200}\frac{u_e^3}{v^2}R_{x_{tr}}^{-1.34}$$

It also provides a computational advantage by allowing the calculations to go smoothly from a laminar flow to a turbulent flow, as shown by the calculations of Schubauer's (1939) elliptic-cylinder flow which are compared with measurements in the panel. Equation (1) is wholly empirical.

6.9 Prediction of transition

Transition may be predicted by:

a. correlations, such as that of Michel (1951),

$$R_\theta = 1{\cdot}174\left[1 + (22400/R_{x_{tr}})\right]R_{x_{tr}}^{0.46}, \quad 10^5 \leq R_x \leq 4 \times 10^7;$$

b. spatial amplification theory based on the solution of the Orr–Sommerfeld equation, "e^9-method"

As implied on the previous two panels, and in the discussion of panel 1.13, there are no exact methods for calculating transition. Present methods use empirical correlations, such as Michel's equation shown on the above panel, or linear stability theory with empirical assumptions.

Michel's correlation, and others like it, must be used within the range of the data upon which it was based. Thus, for example, it should not be used in the presence of wall heat transfer or compressible flow.

The "e^9" method offers the possibility of wider applicability largely because it requires the solution of conservation equations including the stability equation. It has been used, for example, by Smith and Gamberoni (1956) and van Ingen (1956) for two-dimensional flows and for three-dimensional flows by Mack (1977) and by Cebeci and Stewartson (1980a, b). It is discussed further in Chapter 12.

6.10 Importance of wake calculations to lift and drag

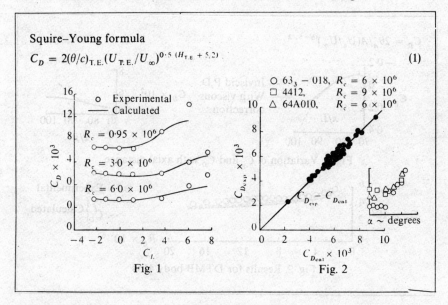

Squire–Young formula
$$C_D = 2(\theta/c)_{T.E.}(U_{T.E.}/U_\infty)^{0.5\,(H_{T.E.}+5.2)} \tag{1}$$

Fig. 1

Fig. 2

The discussion of the previous panels has implicitly assumed that the free-stream pressure distribution is known. For an airfoil or wing the pressure distribution is not known *a priori* but is essential to the correct calculation of local flow properties and overall lift and drag. It can be computed by solving the potential-flow equations for a given geometry, using the resulting pressure distribution on the body for the boundary-layer calculations, solving the potential-flow equations again but for a body shape corresponding to the locus of the displacement thickness, and repeating this sequence of calculations until it converges.

For an accurate calculation of lift and drag by the above iteration procedure, it is essential that the displacement thickness be accurately calculated on the airfoil and in the downstream wake. More will be said of the inviscid–viscid interaction in Chapter 11 and of a procedure for calculating into the wake in the following panels.

Some methods used to calculate the total drag of two-dimensional bodies *avoid* wake calculations and the inviscid–viscid interaction by making use of the so-called Squire–Young formula shown on the panel. This formula makes use of the momentum thickness, shape factor and velocity at the trailing edge; it is limited to incompressible flows although several modified extensions for compressible flows have been reported. Extensive studies by Cebeci and Smith (1974) have shown this formula to be satisfactory for airfoils at zero and at small incidence; it is not satisfactory at high incidence, as shown by the figures of the panel. The satisfactory results of Fig. 2 correspond to 57 drag values at angles of incidence less than 6 degrees.

6.11 Total drag of an axisymmetric body

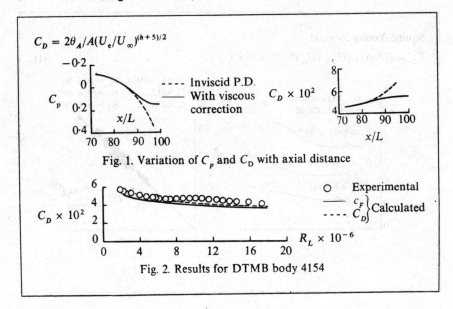

Fig. 1. Variation of C_p and C_D with axial distance

Fig. 2. Results for DTMB body 4154

For bodies of revolution the Squire–Young formula can be modified as shown above. Here h is a shape factor defined as the ratio of displacement area δ_A^* to momentum area θ_A. These are defined by

$$\delta_A^* = 2\pi \int_0^\infty r\left(1 - \frac{U}{U_e}\right) dy, \qquad \theta_A = 2\pi \int_0^\infty r\frac{U}{U_e}\left(1 - \frac{U}{U_e}\right) dy.$$

In the equation for C_D, A is the frontal area based on the maximum radius R_0 of the body and is usually given by $A = \pi R_0^2$.

The prediction of total drag is sensitive to the definition of "trailing edge" because, close to the trailing edge, the inviscid pressure distribution changes drastically from the actual one (see Fig. 1 above) and in some cases the body radius goes to zero.

As long as the pressure drag is not too important, the total drag of an axisymmetric body can be computed satisfactorily from the average skin-friction drag as shown in Fig. 2. Here the trailing edge was taken as the point where $C_p = 0$; for details, see Cebeci et al. (1972b). The calculation of the total drag of an axisymmetric body is not, however, entirely satisfactory and it appears that extending the wall boundary-layer calculations and the inviscid–viscous interaction into the wake is inevitable. For two recent studies of the subject, the reader is referred to Huang et al. (1978) and Patel and Lee (1978). A useful way of calculating into the wake is explained on the following panels.

External Boundary-layer Problems

6.12 A double-structured numerical scheme for wakes, 1

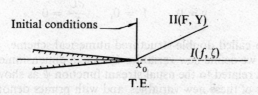

In region II, define new variables Y, s and F by

$$Y = \left(\frac{U_0}{\nu x_0}\right)^{1/2} y, \quad s = \left(\frac{x}{x_0} - 1\right)^{1/3}, \quad F = \frac{\psi}{(U_0 \nu x_0)^{1/2}}$$

and write the thin shear-layer equations and edge boundary condition as

$$3s^2(bF'')' = F'\frac{\partial F'}{\partial s} - F''\frac{\partial F'}{\partial s} - \bar{U}_e \frac{d\bar{U}_e}{ds}$$

and

$$Y \to \infty \quad F' \to \bar{U}_e(s)$$

There are a number of problems which require the solution of an equation or equations with discontinuities in boundary conditions. A two-point finite-difference method like the Box method is ideally suited to solve these problems; the following describes how this can be done by taking the wake calculations as an example.

For simplicity, consider a symmetrical airfoil and assume that the velocity profile is known at the trailing edge, x_0 say from the wall boundary-layer method described previously. A double-structured regime is defined starting at the trailing edge and consisting of an inner region I and an outer region II as shown in the panel. The two-dimensional thin shear-layer equations are again

$$\frac{\partial U}{\partial x} + \frac{\partial V}{\partial y} = 0, \tag{1}$$

$$U\frac{\partial U}{\partial x} + V\frac{\partial U}{\partial y} = U_e \frac{dU_e}{dx} + \frac{\partial}{\partial y}\left(b\frac{\partial U}{\partial y}\right), \tag{2}$$

and for wall boundary-layer flows they are subject to

$$y = 0: \quad U = V = 0, \tag{3a}$$

$$y \to \infty: \quad U \to U_e(x). \tag{3b}$$

Our object is to take equations (1) and (2) and solve them subject to equation (3b) and

$$y = 0, \qquad V = 0, \qquad \frac{\partial U}{\partial y} = 0 \qquad (3c)$$

by using the so-called double-structured numerical scheme.

In region II, we define new variables Y, s and a dimensionless stream function F which is related to the usual stream function ψ as shown in the above panel. In terms of these new variables, and with primes denoting differentiation with respect to Y, the equations reduce to the forms shown on the panel. Here $F' = U/U_0$, $\overline{U}_e = U_e/U_0$ and U_0 and x_0 denote the values of external velocity and surface distance at the trailing edge.

6.13 A double-structured numerical scheme for wakes, 2

In region I, define new variables ζ and f by

$$\zeta \equiv Y/s, \qquad f(s,\zeta) = F/s^2 \tag{4}$$

and write the equations of motion and the boundary conditions as

$$(bf'')' + \frac{2}{3}ff'' - \frac{1}{3}(f')^2 = \frac{s}{3}\left(f'\frac{\partial f'}{\partial s} - f''\frac{\partial f}{\partial s}\right) - \frac{\bar{U}_e}{3s}\frac{d\bar{U}_e}{ds} \tag{5}$$

$$\zeta = 0: \qquad f = f'' = 0 \tag{6}$$

Interface at $\zeta = \zeta^*$:

$$f = \frac{F}{s^2}, \qquad f' = \frac{F'}{s}, \qquad f'' = F'' \tag{7}$$

Similarly in region I, we define new variables ζ and f by equations (4) and rewrite the equations once more, as shown in the panel.

In region I we now let $\zeta \equiv \zeta^*$ denote the position of the interface between Regions I and II, so that the line separating I and II in the (Y,s) plane is given by $Y = \zeta^* s$. At $\zeta = \zeta^*$ we require the continuity of f with F and its derivatives as shown in equation (7). Note that primes on f denote differentiation with respect to ζ and on F denote differentiation with respect to Y.

The solutions of the above equations for regions I and II are obtained by using the Box method as described by Cebeci, Thiele, Williams and Stewartson (1979).

6.14 Turbulence model for symmetrical wakes

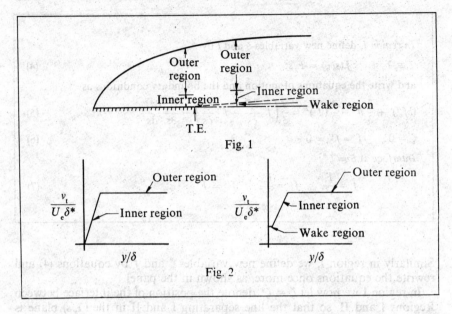

Fig. 1

Fig. 2

The Cebeci–Smith turbulence model of panel 6.7, which has been extensively tested for wall boundary-layer flows, can also be used for near-wake flows, following the ideas set forth by Townsend (1966) and Bradshaw (1970). In the extension of this model to near wake flows, the inner region of eddy viscosity was further subdivided into two parts, as shown above, by Cebeci, Thiele, Williams and Stewartson (1979). Thus we define

$$(v_t)_w = 0.4 \, U_\tau^{T.E.} \, y_c \quad \text{for} \quad 0 < y < y_2, \tag{1}$$

where y_c is the distance from the centreline and

$$U_{calc} = 1.10 \, U_{log}. \tag{2}$$

U_{calc} is the computed velocity profile in the wake, and U_{log} is the logarithmic velocity profile associated with the trailing edge, i.e.

$$U_{log} = U_\tau^{T.E.} \left(2.5 \ln \frac{y U_\tau^{T.E.}}{v} + 5.2 \right). \tag{3}$$

When $y_2 < y < y_1$, we use the inner eddy-viscosity formula of equation (1) in panel 6.7 with $l = 0.4\,y$ and $\gamma_{tr} = 1$. Figures 1 and 2 identify eddy viscosity regions (not to scale) for a wall boundary layer and for a wake flow.

Note that according to the outer eddy viscosity formula

$$(v_t)_0 = 0.0168 \left| \int_0^\infty (U_e - U) \, dy \right| \tag{4}$$

External Boundary-layer Problems

and equation (1) for $(v_t)_w$, the magnitude of $(v_t)_w$ may exceed that of $(v_t)_0$ at large enough distances from the trailing edge. In order to prevent this unacceptable property, the eddy viscosity is taken to be constant and equal to $(v_t)_0$ across the wake for values of x larger than that corresponding to $(v_t)_0 = (v_t)_w$.

6.15 Results for laminar and turbulent near wakes: zero-pressure gradient

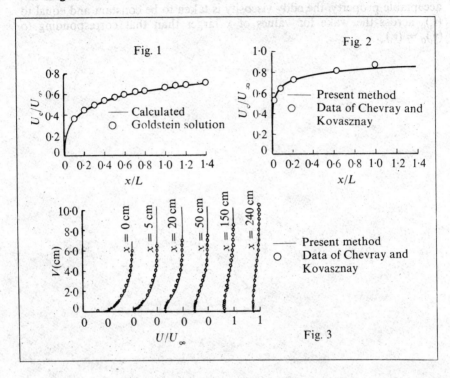

The method which uses the double-structured numerical scheme discussed in previous panels was first applied by Cebeci *et al.* (1979) to laminar wakes with zero pressure gradient to test its numerical accuracy. Figure 1 compares the calculated centreline velocities with those given by Goldstein (see Berger, 1971, p. 12) and, as can be seen, the computed centreline velocities are identical within the precision with which Goldstein's data can be interpreted from the available figure.

Turbulent-flow calculations were then made for a flat plate and the computed results were compared with the experimental data of Chevray and Kovasznay (1969). Figure 2 shows a comparison between the calculated and experimental values of centreline velocity; Figure 3 does the same for the velocity profiles in the wake (including the trailing edge). As can be seen, the agreement indicated in these figures is excellent.

6.16 Separation prediction

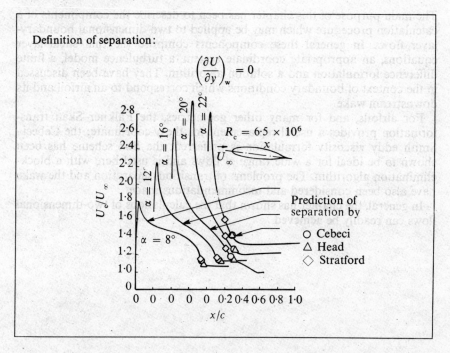

The calculation of separation is important since, once it occurs, the lift drops rapidly and the drag increases. Usually it is required to have a procedure to calculate separation so that a design can be modified to avoid it. In practice the separation point of a two-dimensional flow varies with time but, in common with other properties, has a time-averaged value which may be regarded as corresponding to zero wall-shear stress. Calculation methods, such as that described in the previous pages, are able to predict this point for both laminar and turbulent flows as described by Cebeci et al. (1972a).

The prediction of separation can be achieved in other ways, for example by the solution of integral equations as shown by Head (1958) or by a formula such as that proposed by Stratford (1959). As previously indicated, correlation formulas and integral methods are less widely applicable than differential methods and as shown by the panel which represents the separation on an NACA 66-2-420 airfoil as a function of angle of attack, the correlation equation and the differential method are better than the integral method. It can be expected that the correlation method will be less satisfactory for airfoil shapes which imply large pressure gradients.

This subject is discussed further in Chapter 8 in connection with unsteady two-dimensional and steady three-dimensional boundary layers.

6.17 Concluding remarks

The main purpose of this chapter has been to describe the components of a calculation procedure which may be applied to two-dimensional boundary-layer flows. In general these components comprise the thin shear-layer equations, an appropriate coordinate system, a turbulence model, a finite difference formulation and a solution algorithm. They have been discussed in the context of boundary conditions which correspond to an airfoil and its downstream wake.

For airfoils, and for many other geometries, the Falkner–Skan transformation provides a useful replacement for the y-coordinate; the Cebeci–Smith eddy viscosity formulation is preferred; the Box scheme has been shown to be ideal for a wide range of flows and is used here with a block-elimination algorithm. The problems of transition, separation and the wake have also been considered and recommendations made.

In general, the chapter has shown that the calculations of two-dimensional flows can readily be achieved.

7 Inverse Boundary-layer Problems

7.1 Purposes and outline of chapter

- To describe a procedure for the solution of two-dimensional boundary-layer equations with boundary conditions appropriate to internal flows.
- To introduce the problem of separating boundary layers and small, near-wall, closed regions of separation and to describe a procedure, based on the solution of the boundary layer equations, for calculating their properties

The flow in plane channels and round ducts with uniform cross-section or with cross-sections which do not vary sufficiently to cause separation, are also represented by the boundary-layer equations. The boundary conditions differ from those for an external flow in that the dimensions of the duct are known rather than a freestream velocity variation. This difference requires a change in the numerical method described in the previous chapter and two alternative calculation procedures are described here. In contrast to the "standard" problem of the external boundary layer, internal-flow problems are solved by an "inverse" method where the variation of a centreline velocity is determined from a knowledge of the variation in duct area.

The calculation of mildly separated flows, i.e. those for which the extent of the back flow region is small, may be described by boundary-layer equations; but these become singular at the separation point if the freestream velocity distribution is prescribed. This problem is removed if the displacement-thickness variation, for example, is prescribed instead, and again we have an "inverse" problem. Separating boundary layers and small confined regions of separation can belong to this class of flows and are important in applications, such as wing, airfoil and turbine blades design.

Two inverse boundary-layer methods are described and both make use of the Box method and, for turbulent flow, the Cebeci–Smith algebraic eddy viscosity formulation. The eigenvalue method is preferred for unseparated or internal flows, and the "Mechul"* method for separated flows.

* Pronounced "meshul" (Turkish "unknown").

7.2 Entrance region problem by an inverse boundary-layer method, 1

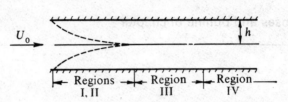

Regions I, II — Region III — Region IV

Displacement interaction region (regions I, II)
Transformation

$$x = x, \quad \eta = \sqrt{U_0/\nu x}\, y, \quad \psi = (U_0 \nu x)^{1/2} f(x, \eta) \quad (1)$$

Governing equation

$$(bf'')' + \tfrac{1}{2} ff'' = x\left(f' \frac{\partial f'}{\partial x} - f'' \frac{\partial f}{\partial x} \right) + x\frac{dP}{dx} \quad (2)$$

Here $f' = U/U_0$ and U_0 is a constant reference velocity: $P = $ (pressure)$/\rho U_0^2$

The internal flow problem is considered first and, as shown in the panel, in the context of a two-dimensional channel of uniform cross-section. The procedures are readily applied to axisymmetric configurations and to variations in cross-section which do not give rise to flow separation.

The duct flow may be divided into four regions. In the first the boundary layers are thin and it is desirable to solve the equations in transformed coordinates, as shown in the panel. The same procedure is used, for convenience, in region II where the boundary layer is thicker and close to the duct half-width (or radius). The flow in the displacement interaction region (I and II) is therefore solved with the differential equation in the same form as those for the external boundary layer. The governing equations in the developing (III) and fully developed region (IV) are solved with equations in physical variables as indicated on panel 7.4.

7.3 Entrance region problem, 2

Boundary conditions for the displacement-interaction region
Wall boundary conditions:

$$\eta = 0, \quad f_0 = f_w(\text{given}), \quad f_0' = 0 \tag{3a}$$

Edge boundary conditions, with $\beta \equiv P_{i-1} - P_i$:

$$\eta = \eta_e, \quad f_e' = \frac{U_e}{U_0} = \sqrt{2\beta + (f_e')_{i-1}^2}. \tag{3b}$$

Conservation of mass:

$$U_0 h = \int_0^h U \, dy$$

$$1 = \frac{x}{h} \frac{f_c}{\sqrt{R_x}}, \quad R_x = \frac{U_0 x}{\nu} \tag{3c}$$

$$f_e = \frac{h}{x}\sqrt{R_x} - f_e'(\eta_c - \eta_e) \tag{3d}$$

For the displacement interaction region, the *wall* boundary conditions are straightforward. The *edge* boundary condition, that is the condition at $\eta = \eta_e, f' = U_e/U_0$, can be written as shown in equation (3b) by making use of Bernoulli's equation. Since U_e (or f' at the edge) is not known, it must be calculated in such a way that the mass balance resulting from the conservation of mass is satisfied. In terms of transformed variables, that relation can be written as shown by equation (3c). Here the subscript c denotes the centreline of the duct. Recalling that the dimensionless stream function f varies linearly with η at the boundary-layer edge, we can write an expression for f_e by making use of (3c) as shown by equation (3d). Here $\eta_c = \sqrt{U_0/\nu x}\, h$.

The equations of this and the previous panel represent the flow in region I and II and provide the basis for solutions of laminar duct flows when the entrance boundary layers have not merged. The turbulence model of panel 7.5 allows this to be extended to turbulent flows.

7.4 Entrance region problem, 3

Governing equations for the shear-layer interaction and fully developed region (regions III, IV)

$$(bF'')' = \frac{dP}{dx} + F'\frac{\partial F'}{\partial x} - F''\frac{\partial F}{\partial x} \qquad (4)$$

Here $F = \psi/(U_0 \nu L)^{1/2}$ with primes denoting differentiation with respect to $Y = (U_0/\nu L)^{1/2} y$

Boundary conditions
Wall
$$Y = 0, \quad F_0 = F_w(\text{given}), \quad F'_0 = 0 \qquad (5a)$$

Centre line
$$Y = \sqrt{R_h}, \quad F''_c = 0, \quad F_c = \sqrt{R_h} \text{ (conservation of mass)} \qquad (5b)$$

Although the similarity variable η removes the singularity at the leading edge (where initial conditions are obtained by the similarity solutions), stretches the coordinate normal to the flow and maintains the transformed boundary-layer thickness nearly constant ($\eta_e \approx 8$), it is necessary to abandon the similarity variable in favour of physical variables at some $x = x_0$ before the end of the displacement interaction region. The governing equations and their boundary conditions in this case are given by equations (4) and (5) in the above panel. Note that in this case we solve the governing equations in the region $0 \leq Y \leq \sqrt{R_h}$, where $R_h = U_0 h/\nu$.

Now the equations for a laminar duct flow have been assembled and can readily be extended to turbulent flow as shown on the next panel. In the fully developed region, of course, equation (4) reduces to an ordinary differential equation and can be solved analytically for laminar flows. It can also be solved with the aid of the Box method and the algorithm described in the following panels.

Inverse Boundary-layer Problems

7.5 Entrance region problem, 4: eddy viscosity formulation

$$v_t = v_t^i + (v_t^f - v_t^i)\{1 - \exp[-(x - x_0)/20 r_0]\} \quad (6)$$

Here subscripts "i" and "f" denote initial and final values, and

$$v_t^i \equiv \begin{cases} = (0\cdot 4\, y)^2 [1 - \exp(-y/A)]^2 \left|\dfrac{\partial U}{\partial y}\right| & (7a) \\[2ex] = 0\cdot 0168 \displaystyle\int_0^\infty (U_e - U)\, dy & \end{cases}$$

$$v_t^f = l^2 \left|\frac{\partial u}{\partial y}\right| \quad (7b)$$

$$l = r_0 \left[0\cdot 14 - 0\cdot 08 \left(1 - \frac{r}{r_0}\right)^2 - 0\cdot 06 \left(1 - \frac{r}{r_0}\right)^4 \right] \quad (7c)$$

The solution of the entrance region equations given in the previous panels requires an eddy viscosity formulation for turbulent flows. The above shows one that has been used successfully by Cebeci and Chang (1978). Note that in equation (6) for the displacement interaction region we use the same eddy viscosity formulation (v_t^i) as for external flows (equation (7)). As before, $A = 26\nu/U_\tau$, $U_\tau = \sqrt{\tau_w/\rho}$. The symbol x_0 denotes the distance where $\delta = r_0$ (where the boundary layers first merge). Note that, for an axisymmetric flow with constant r_0,

$$r = r_0 - y$$

so that

$$1 - \frac{r}{r_0} = \frac{y}{r_0}.$$

To use (6) for plane flows, we simply set

$$r_0 = h,$$

$$1 - \frac{r}{r_0} = \frac{y}{h}.$$

7.6 Eigenvalue method, 1

---

a. Solve the standard problem, that is (2), (3a), (3b) of panels 7.2 and 7.3 for an assumed $\beta (\equiv P_{i-1} - P_i)$

b. Obtain an improved value of β from Newton's method

$$\beta^{\nu+1} = \beta^\nu - \frac{\phi(\beta^\nu)}{\partial \phi(\beta^\nu)/\partial \beta} \qquad \nu = 0, 1, 2, \ldots \qquad (1)$$

(iteration index)

Here

$$\phi \equiv f_e - \frac{h}{x}\sqrt{R_x} + f'_e(\eta_c - \eta_e) = 0 \qquad (2)$$

$$\frac{\partial \phi}{\partial \beta} = \frac{\partial f_e}{\partial \beta} + \frac{\partial f'_e}{\partial \beta}(\eta_c - \eta_e) \qquad (3)$$

Let us consider the governing equations for the displacement interaction region, that is, the system given by equations (2), (3a), (3b) and (3c) of panels 7.2 and 7.3, and discuss a solution procedure for it. Since $\beta (\equiv P_{i-1} - P_i)$ is not known, it is obvious that an iteration procedure is necessary. It can be achieved by treating β either as an eigenvalue or as an unknown function. Here we shall use the former approach and outline the eigenvalue method as discussed by Cebeci and Keller (1974). The latter approach, called the Mechul method, is preferred for the calculation of reversed flows as described in panel 7.11.

In the eigenvalue method we first solve the standard problem and determine ϕ, which represents the residual error. It is clear that from (1), in order to determine the next value of β (i.e. the value on the $(\nu + 1)$th iteration), it is necessary to know $\partial \phi / \partial \beta$ and this can be determined from equation (3), as is discussed in the following two panels.

7.7 Eigenvalue method, 2: the variational equations

> c. The derivatives of f and f' with respect to β are obtained by solving a system of variational equations derived by differentiating the difference equations of the standard problem, namely equations (4a), (4b), (4c) below. If we let
>
> $$F \equiv \frac{\partial f}{\partial \beta}, \quad U^* \equiv \frac{\partial U}{\partial \beta}, \quad g \equiv \frac{\partial G}{\partial \beta}, \qquad (5)$$
>
> then we observe that the form of (4a), (4b) remains unchanged except that $f_j \equiv F_j$, $U_j \equiv U_j^*$, $G \equiv g_j$. Left-hand side of (4c) also remains unchanged but its right-hand side becomes
>
> $$\equiv -2\alpha_i$$

To determine $\partial \phi / \partial \beta$, we differentiate the difference equations of the standard problem with respect to β. To illustrate this aspect of the problem, let us consider equation (2) of panel 7.2 and its boundary conditions as given by equations (3a), (3b), (3c) of panel 7.3. The difference equations, following the step 2 of panel 5.11 for the Box method, with $f' = U$, $U' = G$, are

$$f_j^i - f_{j-1}^i - \frac{h_{j-1}}{2}(U_j^i + U_{j-1}^i) = 0, \qquad (4a)$$

$$U_j^i - U_{j-1}^i - \frac{h_{j-1}}{2}(G_j^i + G_{j-1}^i) = 0, \qquad (4b)$$

$$\frac{(bG)_j^i - (bG)_{j-1}^i}{h_{j-1}} + (\tfrac{1}{2} + \alpha_i)(fG)_{j-\frac{1}{2}}^i$$
$$- \alpha_i[(U^2)_{j-\frac{1}{2}}^i - G_{j-\frac{1}{2}}^{i-1} f_{j-\frac{1}{2}}^{i-1} + f_{j-\frac{1}{2}}^{i-1} G_{j-\frac{1}{2}}^{i-1}] = -2\alpha_i \beta + R_{j-\frac{1}{2}}^{i-1}. \qquad (4c)$$

Here

$$\alpha_i \equiv \frac{x_{i-\frac{1}{2}}}{k_i}, \quad R_{i-\frac{1}{2}}^{i-1} = -\frac{(bG)_j^{i-1} - (bG)_{j-1}^{i-1}}{h_{j-1}} - \tfrac{1}{2}(fG)_{j-\frac{1}{2}}^{i-1} + \alpha_i\{-(U^2)_{j-\frac{1}{2}}^{i-1}$$
$$+ (fG)_{j-\frac{1}{2}}^{i-1}\}.$$

7.8 Eigenvalue method, 3: solution of the variational equations

> Write the variational equations as
> $$A\delta = r \qquad (6)$$
> Where A is same as in standard problem,
> $$\delta_j \equiv \begin{pmatrix} F_j \\ U_j^* \\ G_j \end{pmatrix} \quad \text{and} \quad r_j \equiv \begin{pmatrix} (r_1)_j \\ (r_2)_j \\ (r_3)_j \end{pmatrix} \neq r_j \text{ of the standard problem}$$
>
> Solve (6) to get $\quad F_J \equiv \dfrac{\partial f_J}{\partial \beta}, \quad U_J^* \equiv \dfrac{\partial U_J}{\partial \beta}$
>
> and determine $\partial \phi / \partial \beta$

Note that the variational equations are linear equations. The right-hand side of equation (6) is not the same as the right-hand side of the standard problem. In this case

$$(r_1)_j = 0, \qquad (r_2)_j = -2\alpha_i, \qquad (r_3)_j = 0 \text{ except for } j = J.$$

For $j = J$ differentiate $(U_J^2)^i = 2\beta + (U_J^2)^{i-1}$ to get

$$(r_3)_J = \frac{1}{U_J}.$$

The iteration procedure is repeated until

$$|\beta^{v+1}(x_n) - \beta^v(x_n)| < \gamma_1$$

where γ_1 is a small error tolerance.

7.9 Comparison of calculated and experimental data

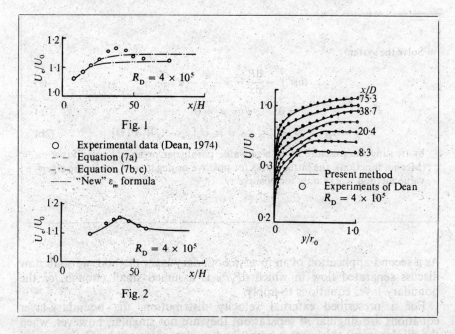

This panel shows a comparison of predictions obtained using the eigenvalue procedure of the previous panels with experimental data of Dean (1974) obtained in plane uniform section channels. Note that when one uses the same eddy viscosity formulation developed for external boundary layers (equation (7a) of panel 7.5) for the four distinct regions of panel 6.13, or when one uses a formulation deduced from fully developed flow, such as equation (7b), (7c), the computed results do not agree well with the experimental data, as shown in Fig. 1. On the other hand, a combination of these formulas as given by equation (6) of panel 7.5 gives very good agreement, as shown in Fig. 2. For further comparisons see Cebeci and Chang (1978).

The number of iterations required to achieve convergence to a satisfactory value of γ_1, i.e. 0·001, was two and the resulting computer times were nearly the same as those required for a similar external boundary-layer calculation; the use of the variational equations does not contribute too much to the calculation time since to solve equation (6) of panel 7.8, we only need to compute r; the coefficient matrix A is known for the standard problem so it does not need to be evaluated again for the variational equations. Also, when we solve for δ defined in panel 7.8, we can stop the calculations of δ_j after $j = J$ in the backward sweep since the calculation of δ_j for all values of j are not needed.

7.10 Separating flows by an inverse boundary-layer method

Solve the system

$$(b\psi'')' - \frac{dP}{dx} = \psi' \frac{\partial \psi'}{\partial x} - \psi'' \frac{\partial \psi}{\partial x} \qquad (1)$$

$$y = 0, \qquad \psi = \psi' = 0 \qquad (2a)$$

$$y = y_e, \qquad \psi' = U_e, \qquad \psi = U_e[y_e - \sqrt{R_L}\delta^*(x)] \qquad (2b)$$

by treating the pressure as an eigenvalue parameter or an unknown function (Mechul function): the latter works for positive or negative τ_w, while the eigenvalue method works only for positive τ_w.

As a second application of an inverse boundary-layer method, we shall now discuss separated flow in which $\partial P/\partial y$ is assumed small enough for the boundary-layer equations to apply.

For a prescribed external velocity distribution, the boundary-layer equations are singular at separation; they are not singular, however, when the displacement thickness is prescribed, as demonstrated by Catherall and Mangler (1966). More recent investigations of external inverse boundary-layer methods have been reported for example, by Klineberg and Steger (1974), Horton (1974), Carter (1974, 1975), Carter and Wornom (1975), Williams (1975), Cebeci and Keller (1972), Keller and Cebeci (1972), Cebeci (1976a, b), Arieli and Murphy (1980) and Cebeci, Keller and Williams (1979).

We now consider the governing equations as given in panel 6.2 and in the above panel which shows the formulation of the problem and two possible approaches to its solution. Note that in equation (2) we specify the displacement thickness δ^*. The Mechul function approach is described in the following pages and, as indicated above, works satisfactorily for forward and reverse flow. The main purpose here is to describe its application to flows with separation. It should be appreciated that, to meet the requirement that $\partial P/\partial y$ is small and the more severe requirement that boundary-layer equations apply, the size of the region of recirculation will be small. Large regions of recirculation must be represented by equations such as those described in Chapter 10.

7.11 Mechul function method: solution procedure, 1

$$\psi' = U \tag{3a}$$
$$U' = G \tag{3b}$$
$$P' = 0 \tag{3c}$$
$$(bG)' - \frac{\partial P}{\partial x} = U\frac{\partial U}{\partial x} - G\frac{\partial \psi}{\partial x} \tag{3d}$$

Boundary conditions

$$y = 0: \psi' = \psi = 0 \tag{4a}$$
$$y = y_e: U = U_e, \psi_e = U_e[y_e - \sqrt{R_l}\delta^*(x)] \tag{4b}$$

In the Mechul function method we solve equations (1) and (2) of panel 7.10 with Keller's Box scheme. With the function P treated as unknown and with the help of the y-momentum equation ($\partial P/\partial y = 0$), equations (1) and (2) may be written as shown above, in a similar way to those in panel 6.2. The relation between P and U_e is given by Bernoulli's equation, which can be written as in panel 7.3,

$$P^i + \frac{(U_e^2)^i}{2} = P^{i-1} + \frac{(U_e^2)^{i-1}}{2}.$$

We next write the difference equations for these equations and linearize them as described in panels 6.2 through 6.4. There are now four unknowns rather than three and the linearized system can be solved by the block elimination method, now with 4×4 blocks.

This procedure can also be used instead of the eigenvalue method for internal-flow problems. Here it is reserved for application to inverse problems where separated flow is involved and the following pages describe its extension to such problems.

7.12 Mechul function method: solution procedure, 2

FLARE approximation
Neglect $U(\partial U/\partial x)$ in the region of negative U-velocity.

DUIT procedure (*due to Williams*, 1975)
Sweep 1: A solution is generated in the bubble by the FLARE approximation.
Sweep 2: Start at x_f and march upstream in the region of negative wall shear with the range of integration as shown by dashed lines. Solve the regular boundary-layer equations for the given "edge" velocity distribution.

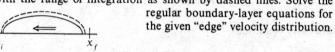

Sweep 3: Start at x_i and march downstream using the inverse boundary-layer procedure again but retaining the $U(\partial U/\partial x)$ term.

For flows with negative wall shear it is necessary to overcome the stability problem to continue the calculations past the separation point. This may be done with the approximation suggested by Reyhner and Flügge-Lotz (1968). This approximation, referred to as FLARE by Williams (1975), consists of the neglect of the $U\partial U/\partial x$ term as indicated in the panel.

Once a solution is generated by the FLARE approximation, more accurate solutions of the governing equations can be obtained in several ways. The best approach, at the moment, is that involving the DUIT procedure due to Williams (1975). This procedure consists of an upstream sweep from x_f to x_i (sweep 2 in the panel) in which the standard problem is solved with the "edge" velocity distribution outside the bubble computed in sweep 1. In the downstream sweep (sweep 3 in the above panel), we again use the inverse procedure, but this time the term $U\partial U/\partial x$ is not neglected. In the region where U is negative, this term is computed from the values of U obtained in sweep 2. When U is positive, it is computed in the normal way. This procedure is then repeated until convergence. According to the investigations of Cebeci, Keller and Williams (1979), this procedure converges in 3 to 4 complete sweeps.

7.13 Results for laminar flows

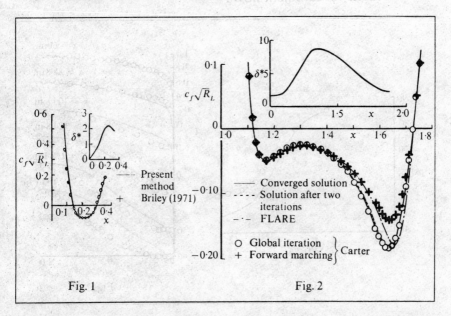

Fig. 1

Fig. 2

Figure 1 of the panel shows comparisons of local skin-friction coefficient $c_f\sqrt{R_L}$ calculated with the Mechul method with those obtained by Briley (1971) from the steady two-dimensional form of the Navier–Stokes equations. These calculations were made for the $\delta^*(x)$ distribution deduced from the Navier–Stokes solutions. The small discrepancy may be associated with the difficulty of reading the input $\delta^*(x)$ distribution from the graph presented by Briley. The Navier–Stokes solutions were obtained for $R_L = 2.08 \times 10^4$ and required 45 minutes on a UNIVAC 1108. The results obtained by the Mechul function method with FLARE required approximately 10 seconds on a CDC 6600 computer.

Figure 2 of the panel allows comparison of calculated results obtained by Carter (1975) and with the Mechul method, and shows that the present results agree well with those of Carter. The computer time was less than 10 seconds for one sweep on a CDC 6600 for a fine grid of over 21 000 points ($\Delta x = 0.005$, $\Delta y = 0.1$). At all x-stations, including regions of separated flow, the convergence was quadratic and required only two to four iterations. The observed error at the termination of the iterations was 10^{-8}. For this case, Carter required 166 iterations, leading inevitably to much higher computer times.

These results show the great advantage, in simplicity and speed, of using the boundary-layer equations if they are applicable, as in the case of shallow separation bubbles.

7.14 Results for turbulent flows

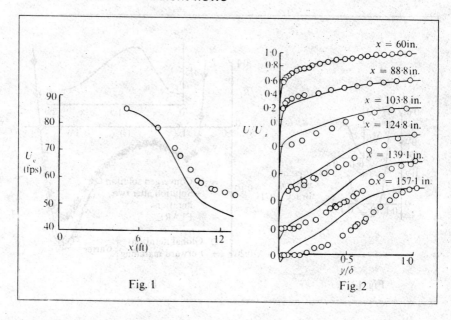

Fig. 1

Fig. 2

To test the Mechul method for separated turbulent flows, we performed calculations for the experimental data of Simpson et al. (1974, 1977). The boundary-layer calculations were first made in the form of the standard problem with the measured freestream velocity distribution and the velocity profile measured at 60 inches. No region of reverse flow was detected and the calculations proceeded to the end of the flow, with 50 grid points across the layer, in 10 seconds. The calculation was repeated, this time by using the Mechul method with the given experimental displacement-thickness distribution. This inverse calculation did encounter a region of separation flow, with the separation point in accord with the experiment, and allowed the determination of the freestream velocity distribution which is shown in Fig. 1 of the above panel; the corresponding nondimensional profiles are shown in Fig. 2. These results are described in greater detail by Cebeci, Khalil and Whitelaw (1979).

Similar calculations using a different numerical method and three turbulence models (algebraic eddy viscosity or mixing length) have been performed by Pletcher (1978).

7.15 Concluding remarks

This chapter has described two numerical methods for solving inverse boundary-layer problems. Both have made use of the Box scheme and the same solution algorithm, with a small number of added iterations, as previously described for the standard problem.

The eigenvalue method is recommended for the solution of internal-flow problems. It has been applied here to plane and axisymmetric ducts of uniform cross-section but can be used with similar ease to determine the flow in ducts of varying cross-sections.

The Mechul-function method can also be used for flows with negative velocity and has been applied here to a laminar separation bubble and to a separating turbulent boundary layer. For these calculations, the displacement thickness has been specified and was obtained in one case from a solution of the Navier–Stokes equations and in the other from measurement. It remains to determine the precision with which calculations can be made by iteration between a potential-flow procedure and the Mechul-function method. Of course, the range of separated flows for which the boundary-layer equations are applicable is necessarily limited but, where they can be used, the savings in computer costs are large.

8 Unsteady Two-dimensional and Steady Three-dimensional Flows

8.1 Purposes and outline of chapter

> To describe procedures to solve the problems of:
> - two-dimensional unsteady boundary-layer flow;
> - three-dimensional steady boundary-layer flow, with and without circumferential-flow reversal
>
> An impulsively started cylinder and a fixed prolate spheroid are taken as examples of the two problems, respectively

A major current problem in boundary-layer theory is to include regions of reverse crossflow or of backflow in the boundary layer. In two dimensions the problem is associated with separation and leads to three difficulties:

1. The possible appearance of a singularity at the point of zero skin friction.
2. Numerical instabilities resulting from integration opposed to the flow direction.
3. Rapid thickening of the boundary layer beyond separation.

In three dimensions reverse crossflow is common in flows over bodies of revolution, wings, ship hulls, etc., even though the flow remains attached in the generally accepted terminology

In Chapter 7 we discussed the calculation of two-dimensional steady boundary layers with separation. In this chapter and in Chapters 9 and 13 we shall present and discuss numerical procedures for computing two-dimensional unsteady flows with backflow and three-dimensional steady flows with reverse crossflow and study the separation problem in these flows.

In both cases the numerical schemes are explained and results presented. Further detail can be found in the papers of Cebeci (1979), Cebeci and Carr (1978, 1980) and Cebeci, Carr and Bradshaw (1979).

8.2 Two-dimensional unsteady thin shear-layer equations, 1

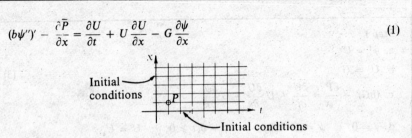

$$(b\psi'')' - \frac{\partial \bar{P}}{\partial x} = \frac{\partial U}{\partial t} + U\frac{\partial U}{\partial x} - G\frac{\partial \psi}{\partial x} \quad (1)$$

Boundary conditions (no mass transfer)

$$\begin{aligned} y = 0, & \quad \psi = \psi' = 0 \\ y = \delta, & \quad \psi' = U_e(x, t) \end{aligned} \quad (2)$$

and initial conditions as shown in the diagram

The unsteady, two-dimensional boundary-layer equations for an incompressible flow may be written in the form ($\bar{P} = P/\rho, b = v + v_t$)

$$\frac{\partial U}{\partial x} + \frac{\partial V}{\partial y} = 0,$$

$$\frac{\partial U}{\partial t} + U\frac{\partial U}{\partial x} + V\frac{\partial U}{\partial y} = -\frac{\partial \bar{P}}{\partial x} + \frac{\partial}{\partial y}\left(b\frac{\partial U}{\partial y}\right),$$

and may be compared with the steady, three-dimensional equations of panel 8.13. As can be seen, the forms of the equations considered are both parabolic and both can readily be solved with the help of the Box method. The procedure for the above equations is described in panels 8.3 and 8.4.

The equations and their boundary conditions may be rewritten in terms of the stream function as shown by equations (1) and (2) of the panel. The boundary conditions are assumed to have the form shown. We now have the usual no-slip wall-boundary condition, and the usual "edge" condition. We also need initial conditions at $x = 0$ in the (y, t) plane and initial conditions at $t = 0$ in the (x, y) plane.

8.3 Two-dimensional unsteady thin shear-layer equations, 2: Box method

Step 1

a. $\psi' = U$

b. $U' = G$

c. $(bG)' - \dfrac{\partial \bar{P}}{\partial x} = \dfrac{\partial U}{\partial t} + U\dfrac{\partial U}{\partial x} - G\dfrac{\partial \psi}{\partial x}$ (3)

d. $y = 0, \quad \psi = U = 0; \qquad y = \delta, t \geqslant 0; \qquad U = U_e$

Step 2
Write the difference equations for (3) for the net cube

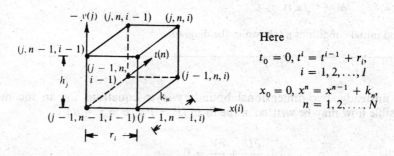

Here
$t_0 = 0, \; t^i = t^{i-1} + r_i,$
$i = 1, 2, \ldots, I$
$x_0 = 0, \; x^n = x^{n-1} + k_n,$
$n = 1, 2, \ldots, N$

The procedure for solving the equations of the previous panel is the same as that described in Chapters 5 and 6. Following the procedure of step 1 of panel 5.11 the definitions of equations (3a, b) allow the differential equation (1) of panel 8.2 to be written as (3c) with the boundary conditions of (3d). This first-order system of equations can then be written in finite difference form for the cube, as shown in the panel and discussed in the following panel.

The important difference for the time-dependent problem, with its third independent variable, is that the basic cell is a cube rather than a square. We now have k, h and r corresponding to increments of x, y and t.

8.4 Two-dimensional unsteady thin shear-layer equations, 3: Box method (continued)

Step 2 (continued)
Centre (3a, b) on $(x^n, t^i, n_{j-\frac{1}{2}})$
Model equation for (3c)

$$(bG)' = \frac{\partial U}{\partial t} + U \frac{\partial U}{\partial x} \qquad (4)$$

Difference approximations to (4) are

$$\frac{(\overline{bG})_j - (\overline{bG})_{j-1}}{h_{j-1}} = \frac{\overline{U}_i - \overline{U}_{i-1}}{r_i} + \overline{U}_{j-\frac{1}{2}} \frac{\overline{U}_n - \overline{U}_{n-1}}{k_n} \qquad (5)$$

Follow *steps* 3 *and* 4 as on panel 5.10, i.e. linearize and solve

---

We write the difference equations for (3a, b) as before by centring them for the midpoint $\eta_{j-\frac{1}{2}}$ at x^n, t^i; that is,

$$\frac{\psi_j^{n,i} - \psi_{j-1}^{n,i}}{h_{j-1}} = U_{j-\frac{1}{2}}^{n,i}, \qquad \frac{U_j^{n,i} - U_{j-1}^{n,i}}{h_{j-1}} = G_{j-\frac{1}{2}}^{n,i}.$$

The difference equations that approximate (3c) are rather lengthy so, for convenience of presentation, the terms $G(\partial \psi / \partial x)$ and $\partial \overline{P}/\partial x$ of equation (3c) have been omitted; the reduced equation, given by equation (4) of the above panel, is then approximated by equation (5), where

$$\overline{G}_j = \tfrac{1}{4}[G_j^{n,i} + G_j^{n,i-1} + G_j^{n-1,i-1} + G_j^{n-1,i}]$$
$$\ldots\ldots\ldots\ldots\ldots\ldots,$$
$$\overline{U}_n = \tfrac{1}{4}[U_j^{n,i} + U_j^{n,i-1} + U_{j-1}^{n,i} + U_{j-1}^{n,i-1}]$$
$$\ldots\ldots\ldots\ldots\ldots\ldots,$$
$$\overline{U}_i = \tfrac{1}{4}[U_j^{n,i} + U_j^{n-1,i} + U_{j-1}^{n,i} + U_{j-1}^{n-1,i}]$$
$$\ldots\ldots\ldots\ldots\ldots\ldots$$

The terms indicated by dotted lines correspond to known quantities.

8.5 Sample calculation, 1: laminar flow over a circular cylinder started impulsively from rest: formulation

---

Small time

$$\eta = \frac{y}{(vt)^{1/2}}, \qquad \psi = (vt)^{1/2} U_e(x) f(x, \eta, t) \qquad (1)$$

$$f''' + \frac{\eta}{2} f'' + t \frac{dU_e}{dx}[1 - (f')^2 + ff''] = t\left[\frac{\partial f'}{\partial t} + U_e\left(f' \frac{\partial f'}{\partial x} - f'' \frac{\partial f}{\partial x}\right)\right] \qquad (2)$$

$$\eta = 0, \quad f = f' = 0; \qquad \eta = \eta_e, \qquad f' = 1 \qquad (3)$$

Initial conditions

In the (x, η) plane for $t = 0$, use the Rayleigh solution. In the (t, η) plane at $x = 0$ and 1, use equation (2) with $U_e = 0$ and $dU_e/dx = 1$ for $x = 0$ and -1 for $x = 1$

As an example of an unsteady flow we consider the problem of laminar flow over a circular cylinder started impulsively from rest (see Cebeci, 1979). Many investigators have computed this flow, both by the thin shear-layer equations and by the Navier–Stokes equations. It should be noted that after a certain time, the streamwise velocity profile contains regions of backflow. The computational problem may be viewed as one step more complex than the problem of two-dimensional steady flows; it is an excellent example for demonstrating the extension of the computational procedures and serves as an introduction to the yet more complex ones to be used for steady three-dimensional flows in which the transverse flow contains regions of backflow. For simplicity we use the inviscid velocity distribution, thus avoiding the problem of the viscid/inviscid interaction discussed in Chapter 11.

Starting the flow impulsively from rest leads to a singularity at $t = 0$. This can be avoided by using an appropriate similarity variable η and by the introduction of a dimensionless stream function $f(x, \eta, t)$ both defined by equation (1) in the above panel. In terms of this transformation, it can be shown that the continuity and momentum equations, together with their boundary and initial conditions, for an impulsively started circular cylinder can be written in the form given by equations (1) and (2). Note that in the (x, η) plane for $t = 0$, the solutions of

$$f''' + \tfrac{1}{2}\eta f'' = 0$$

are given by

$$f = \eta \, \text{erf}(\eta/2) + \frac{2}{\pi^{1/2}}[\exp(-\eta^2/4) - 1]$$

The variables defined by equation (1) are employed for some interval

$0 \leq t \leq t^*$ during which the boundary layer rapidly develops. For $t \geq t_*$ (large time), we switch to other dimensionless variables Y and F defined by

$$Y = \frac{y}{L}, \qquad \psi = U_e L F(x, Y, t) \tag{4}$$

and write the continuity and momentum equations and their boundary conditions in the form

$$F''' + \frac{dU_e}{dx}[1 - (F')^2 + FF''] = \frac{\partial F'}{\partial t} + U_e\left[F'\frac{\partial F'}{\partial x} - F''\frac{\partial F}{\partial x}\right], \tag{5}$$

$$Y = 0, \quad F = F' = 0; \qquad Y = Y_e, \quad F' = 1. \tag{6}$$

Here L is a reference length, $\pi d/2$, x, t, and U_e are nondimensional quantities defined by x/L, $U_\infty t/L$, and U_e/U_∞, respectively, and the primes denote differentiation with respect to Y. The initial conditions in the (t, Y) plane at $x = 0$ and $x = 1$ are obtained from

$$F''' + \frac{dU_e}{dx}[1 - (F')^2 + FF''] = \frac{\partial F'}{dt}$$

with dU_e/dx being equal to 1 for $x = 0$ and -1 for $x = 1$.

8.6 Sample calculation, 2: zig-zag Box

If $U_{j-\frac{1}{2}}^{i,\,n} \geq 0$, then use the standard Box method. If $U_{j-\frac{1}{2}}^{i,\,n} < 0$, write equation (7c) for the Box centred at P using quantities centred at P, Q, R, where

$$P \equiv (x_i, Y_{j-\frac{1}{2}}, t_{n-\frac{1}{2}}), \qquad Q \equiv (x_{i-\frac{1}{2}}, Y_{j-\frac{1}{2}}, t_n), \qquad R \equiv (x_{i+\frac{1}{2}}, Y_{j-\frac{1}{2}}, t_{n-1})$$

We choose t_* small enough so that when we solve equation (2) there is no flow reversal across the boundary layer. The numerical procedure for solving equations (2) and (5) for the case of no flow reversal is quite straightforward. Their solutions subject to their boundary and initial conditions can be obtained by the regular Box method described in panels 8.3 and 8.4. For example, to solve equation (5), we define new dependent variables and write

$$F' = U, \tag{7a}$$

$$U' = G, \tag{7b}$$

$$G' + \frac{dU_e}{dx}(1 - U^2 + FG) = \frac{\partial U}{\partial t} + U_e\left(U\frac{\partial U}{\partial x} - G\frac{\partial F}{\partial x}\right). \tag{7c}$$

In order to compute flows that contain regions of backflow, it is obvious that some upstream influence must be allowed when the streamwise velocity changes sign. Here we use an adaptation of the zig-zag differencing introduced by Krause *et al.* (1969) and also employed by Telionis and Tsahalis (1973) for a similar purpose. However, we do it in a way that maintains a very compact difference scheme, as shown in the panel above.

In the zig-zag scheme we use the normal procedure for equations (7a) and (7b) and centre them at $(x^n, t^i, Y_{j-\frac{1}{2}})$. To write the difference approximation for (7c), centred at $(x^{n-\frac{1}{2}}, t^{i-\frac{1}{2}}, Y_{j-\frac{1}{2}})$, we examine computed values of $U_{j-\frac{1}{2}}^{i,\,n}$ and use the procedure described in the above panel.

8.7 Sample calculation, 3: zig-zag Box (continued)

$$G'(P) + \frac{dU_e}{dx}(P)[1 - U^2(P) + FG(P)] = \frac{\partial U}{\partial t}(P)$$
$$+ U_e(P)\left\{\theta U(Q)\frac{\partial U}{\partial x}(Q) + \phi U(R)\frac{\partial U}{\partial x}(R) - \left[\theta G(Q)\frac{\partial F}{\partial x}(Q) + \phi G(R)\frac{\partial F}{\partial x}(R)\right]\right\}$$

Here
$$\theta = \frac{x_{i+1} - x_i}{x_{i+1} - x_{i-1}}, \qquad \phi = \frac{x_i - x_{i-1}}{x_{i+1} - x_{i-1}} \tag{9}$$

To illustrate the centring of equation (7c) with this zig-zag scheme, we write equation (7c) as shown in the above panel.

The difference equations which are to approximate equation (8) are

$$h_j^{-1}(G_j^{i,n-\frac{1}{2}} - G_{j-1}^{i,n-\frac{1}{2}}) + \left(\frac{dU_e}{dx}\right)^i \left[1 - (U^2)_{j-\frac{1}{2}}^{i,n-\frac{1}{2}} + (FG)_{j-\frac{1}{2}}^{i,n-\frac{1}{2}}\right]$$
$$= k_n^{-1}(U_{j-\frac{1}{2}}^{i,n} - U_{j-\frac{1}{2}}^{i,n-1}) + U_e^i \tilde{\theta}[U_{j-\frac{1}{2}}^{i-\frac{1}{2},n}(U_{j-\frac{1}{2}}^{i,n} - U_{j-\frac{1}{2}}^{i-1,n})$$
$$- G_{j-\frac{1}{2}}^{i-\frac{1}{2},n}(F_{j-\frac{1}{2}}^{i,n} - F_{j-\frac{1}{2}}^{i-1,n})] + \beta_2.$$

Here
$$\tilde{\theta} = \frac{r_{i+1}}{r_i} \frac{1}{(x_{i+1} - x_{i-1})},$$
$$\beta_2 = -\frac{U_e^i r_i}{(x_{i+1} - x_{i-1})^2}[U_{j-\frac{1}{2}}^{i+\frac{1}{2},n-1}(U_{j-\frac{1}{2}}^{i,n-1} - U_{j-\frac{1}{2}}^{i+1,n-1})$$
$$- G_{j-\frac{1}{2}}^{i+\frac{1}{2},n-1}(F_{j-\frac{1}{2}}^{i,n-1} - F_{j-\frac{1}{2}}^{i+1,n-1})].$$

The resulting algebraic system is again nonlinear and its solution is obtained by using the procedure followed in the standard Box method. For further details see Chapter 9.

8.8 Results up to zero skin friction point

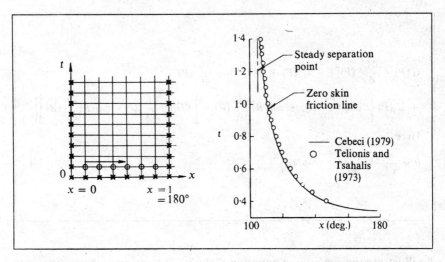

According to the calculations of Cebeci (1979) and others, at approximately $t = 0.33$ the skin friction coefficient vanishes at $x = 1$ (180 degrees). As time increases, the x-location corresponding to zero skin friction decreases from $180°$ and approaches its steady-state value of $1.82 (\equiv 105°)$ as shown in the above panel. Note that the calculations obtained with the present method, indicated by the solid line, are in good agreement with those obtained by Belcher et al. (1971) and Telionis and Tsahalis (1973).

In unsteady flows, unlike steady flows, the vanishing of wall shear (skin friction) does not correspond to flow separation and it is possible to carry the calculations past zero skin-friction point by taking proper account of the backflow. The manner in which this is done is quite important and can lead to erroneous conclusions about flow separation in time-dependent flows. The strict definition of separation is a departure of the shear layer from the surface, i.e. $d\delta/dx$ becoming of order unity.

8.9 Results beyond zero skin friction point, 1

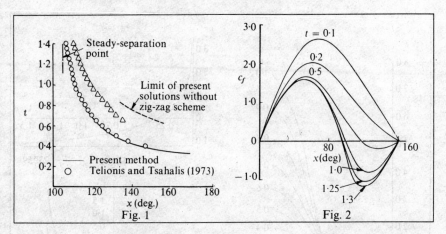

Fig. 1

Fig. 2

The numerical method used by Telionis and Tsahalis was modified to account for the backflow. Results were obtained by this method for the points past zero skin friction. They observed that for $t < 1$, their predictions agreed well with the Navier–Stokes solutions of Thoman and Szewczyk (1966) for solutions up to zero c_f. For $t > 1.0$ they could not obtain solutions. From $t \simeq 0.35$ to $t \simeq 0.65$, their numerical method was able to compute the flow field up to $x \cong 180°$. At about $t = 0.65$, and in the neighborhood of $x = 140°$, a singularity appeared and they were unable to obtain solutions for a few more x-stations beyond the one that corresponds to zero c_f.

Cebeci's calculations (1979) showed, however, that with the standard (unmodified) Box method, the flow field for all x up to and including $t = 0.60$ can be calculated. At the next time interval, $t = 0.65$, calculations were performed without any signs of trouble up to and including $x = 152.4°$ (see Fig. 1 above). Soon thereafter, both in time and in x, it become necessary to use the zig-zag Box, which enabled the computations to continue without difficulty for much longer times, as shown in Fig. 2 in the above panel. His results agreed well with those obtained by Belcher et al. (1971) but did not agree with those obtained by Telionis and Tsahalis (1973) for $t > 0.65$.

The above comparison of the capabilities of two methods for unsteady two-dimensional boundary-layer flow is important in that it shows that different numerical approaches can lead to significantly different capabilities. The extra effort expended in deriving a consistent numerical scheme can reap useful rewards.

8.10 Results beyond zero skin friction point, 2

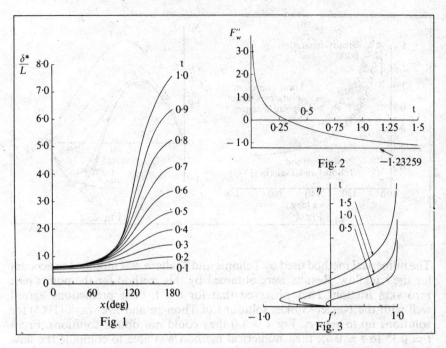

Fig. 1

Fig. 2

Fig. 3

Figure 1 in the above panel shows the dimensionless displacement thickness δ^*/L around the circular cylinder for different values of t ranging from 0·1 to 1·0. Belcher et al. (1971) also computed the δ^*/L values for $t \leqslant 0.70$ for all x and for $t = 1.0$ for x up to $130°$ and comparison with Cebeci's computed δ^* values shows reasonable argument.

Figures 2 and 3 show the results for the rear stagnation point in order to point out the rapid growth (thickening) of the boundary layer with increasing time. Figure 2 shows the variation of the wall-shear parameter F''_w, which appears to agree well with the predictions of Proudman and Johnson (1962) in that the slope of the wall shear parameter at $x = 0$ should tend to the same value as the slope at $x = 1$ as t becomes large.

Figure 3 shows the computed velocity profiles for various values of t. As can be seen, the region of backflow is quite small at small values of x and t (as expected) and becomes quite large with increasing x and t.

8.11 Difficulties with three-dimensional boundary-layer calculations on bodies of revolution at incidence, 1

- *Nose region*: With increasing α, flow may separate on the leeside close to the nose. For laminar flow there is a critical value α such that for $\alpha < \alpha_s$, separation occurs on the leeside near the tail, but for $\alpha > \alpha_s$, separation occurs very near the nose, the change taking place in a dramatic fashion
- *Turbulence modelling*: Extension of 2-D turbulence models to 3-D flows requires additional verification by experiment
- *Negative circumferential-flow velocity*. Depending on the flow conditions, W may contain regions of flow reversal. Such changes in W-profiles can lead to numerical instabilities resulting from integration opposed to the flow direction unless appropriate changes are made in the solution procedure

Calculation of three-dimensional boundary layers on bodies of revolution at incidence is an excellent vehicle for discussing the computational difficulties associated with three-dimensional boundary layers. There are several new features associated with this problem; we describe three of the more important ones as indicated in the above panel. As an example of a body of revolution, we shall consider a prolate spheroid with thickness t. This is a convenient shape to study since analytical expressions are available for its inviscid pressure distribution and we can, therefore, avoid using inviscid computer codes to predict the inviscid flow.

The first comment of the panel indicates that the flow configuration is complex, particularly in the nose region, and is strongly dependent on the angle of attack as shown in the following panel. The nose and downstream regions are discussed separately in the following pages.

So far, we have considered two-dimensional boundary-layer flows and have made use of the algebraic eddy viscosity hypothesis of Cebeci and Smith. In the present three-dimensional flow, however, an appropriately extended turbulence model is required with corresponding experimental verification. This topic is not considered in this chapter but an appropriate form of the eddy viscosity formulation is given in panel 13.4. More sophisticated models can also be used and have been discussed in Chapters 3 and 4.

The emphasis of the following panels is on the solution of numerical difficulties such as those mentioned in the third paragraph of the above panel.

8.12 Difficulties with three-dimensional boundary-layer calculations on bodies of revolution at incidence, 2

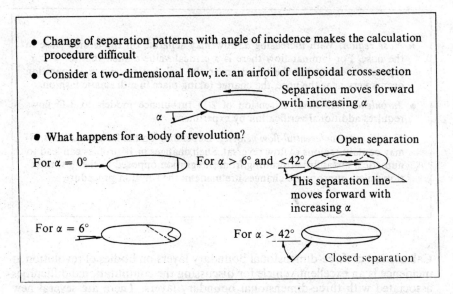

- Change of separation patterns with angle of incidence makes the calculation procedure difficult
- Consider a two-dimensional flow, i.e. an airfoil of ellipsoidal cross-section
- What happens for a body of revolution?

There are several reasons why it is important to develop boundary layer codes for three-dimensional flows on bodies of revolution. Two main ones are:

1. Prediction of aerodynamic forces and flow separation at small and large angles of attack. Here it is desirable to solve the governing equations and study the separation pattern on the leeside and downstream of the body with increasing angle of attack. It is also desirable to compute viscous flows up to separation and analyse their nature (we shall discuss these points later in some detail).
2. Nose transition. In a number of applications it is desirable to predict transition on bodies of revolution at incidence. Due to its very complex nature, adequate theoretical methods are not available for predicting transition and recourse must be made to methods that contain some empiricism. The so-called e^n-method (to be discussed in Chapter 12) is the only promising method for predicting transition on bodies of revolution. It makes use of the solutions of the boundary-layer equations and the linear stability equations. Thus, for proper use of this method, an accurate calculation of three-dimensional boundary layers is needed.

In addition to the difficulties outlined in panel 8.11, change of separation patterns with angle of incidence makes the calculation procedure more difficult, as shown in the above sketches. Here the separation patterns on prolate spheroids are for laminar flows and are sketched according to the calculations of Wang (1976).

8.13 Formulation for a prolate spheroid: governing equations and their boundary and initial conditions

$$\frac{\partial}{\partial x}(h_2 U) + \frac{\partial}{\partial \theta}(h_1 W) + \frac{\partial}{\partial y}(h_1 h_2 V) = 0 \tag{1}$$

$$\frac{U}{h_1}\frac{\partial U}{\partial x} + \frac{W}{h_2}\frac{\partial W}{\partial \theta} + V\frac{\partial W}{\partial y} - W^2 K_2 = -\frac{1}{\rho h_1}\frac{\partial P}{\partial x} + \frac{\partial}{\partial x}\left(v\frac{\partial U}{\partial y} - \overline{uv}\right) \tag{2}$$

$$\frac{U}{h_1}\frac{\partial W}{\partial x} + \frac{W}{h_2}\frac{\partial W}{\partial \theta} + V\frac{\partial W}{\partial y} - UWK_2 = -\frac{1}{\rho h_2}\frac{\partial P}{\partial \theta} + \frac{\partial}{\partial y}\left(v\frac{\partial W}{\partial y} - \overline{vw}\right) \tag{3}$$

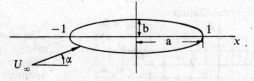

For a prolate spheroid at incidence, the three-dimensional boundary-layer equations for incompressible laminar and turbulent flows in a curvilinear orthogonal coordinate systems are given by equations (1) to (3) in the above panel. Here h_1, h_2, K_2 are geometric parameters of the coordinate system and are given by analytic formulas as a result of our choice of a simple body. Also the external velocity distribution is given by simple analytical formulas for the same reason, and the pressure gradients in equations (2) and (3) are easily computed from them.

The solution of equations (1) to (3) requires boundary and initial conditions. The boundary conditions are:

$$y = 0, \quad U = V = W = 0, \tag{4}$$
$$y \to \infty, \quad U \to U_e(x, \theta), \quad W \to W_e(x, \theta).$$

The initial conditions in the longitudinal direction can easily be determined by taking advantage of the symmetry conditions (see, for example, Cebeci, Khattab and Stewartson (1980a). The specification of the initial conditions in the circumferential direction is not quite so easy when body-oriented coordinates are used because of the singularity of h_1, h_2 and K_2 at the nose ($x = -1$). We shall refer to this problem as the nose-region problem and discuss it in the next panel.

8.14 Nose-region problem and its calculation procedure

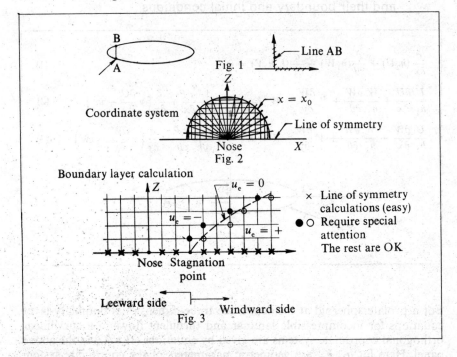

For small angles of incidence, the stagnation point A is very close to the nose ($x = -1$); if $\alpha = 6°$, for example, $x = -0.99898$, but for $\alpha = 30°$, $x = -0.97055$. Therefore, at small angles of attack the assumptions close to the nose, along the line AB (see panel), are not important since separation occurs far downstream where the effects of the nose region have disappeared. At high incidence (i.e. $\alpha > 42°$) separation occurs near the nose, upstream of the line AB; this implies that the nose region is very important at larger angles of attack. For stability and transition calculations, the nose region is very important for all α.

To avoid the difficulties associated with the nose region, Cebeci, Khattab and Stewartson (1980a) introduced a transformation which eliminated the geometric singularity at the nose by a special transformation $(x, \theta) \Rightarrow (X, Z)$ which essentially converts polar coordinates to quasi-rectangular Cartesian form. This transformation allows the initial conditions in the circumferential direction to be successfully generated. Since the expressions resulting from the transformation are rather lengthy, they are not reproduced here.

Figs 2 and 3 in the above panel show the coordinate system and the procedure used for the boundary-layer calculations in the nose region. We first calculate the line of symmetry solutions and then successive stations are computed using the characteristic Box (to be discussed in Chapter 13) until $x = x_0$ when we switch to the general body coordinates (x, θ, y).

8.15 Boundary-layer calculations away from the nose

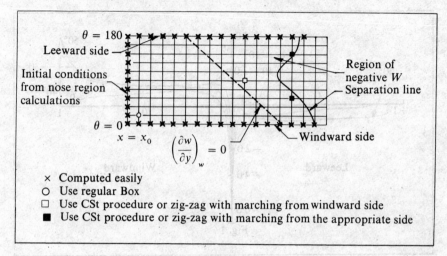

× Computed easily
○ Use regular Box
□ Use CSt procedure or zig-zag with marching from windward side
■ Use CSt procedure or zig-zag with marching from the appropriate side

Similarity type transformations are applied to the governing equations expressed in the (x, θ, y) coordinates which are solved as described in Cebeci, Khattab and Stewartson (1980a).

The numerical solution procedure used to perform the boundary layer calculations away from the nose depends on the complexity of the flow as shown in the above panel and below. Let us assume that the initial conditions in the circumferential direction and along the windward and leeward sides are obtained by using the regular Box method. These solutions are indicated by x's in the above panel. In the regular Box method, the governing equations are centred as shown below. It can be used with accuracy as long as the circumferential flow velocity W is positive across the layer.

The region to the right of the dashed line shown in the panel corresponds to the region where W becomes negative. In the part of this region between the dashed line and the separation line either the procedure of Cebeci and Stewartson (1978) or the zig-zag procedure, both indicated below, should be used with the solutions marching from the windward side. This is indicated by the symbol □. If, as the solutions progress in the windward direction, the separation line is encountered, we change the marching direction to approach from leeward. This is indicated by the symbol ■.

The regular Box and the Cebeci–Stewartson procedure, which is the Characteristic Box, will be discussed further in Chapter 13. The finite difference patterns are:

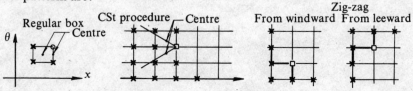

8.16 Results for line of symmetry; finite spheroids

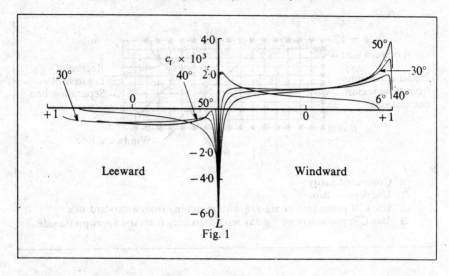

Fig. 1

Figure 1 shows the variation of the longitudinal local skin friction coefficient c_f for various angles of incidence and thickness ratio $t = \frac{1}{4}$. These results agree well with those computed by Hirsh and Cebeci (1977) who considered only small angles of incidence, and with those of Wang (1970) who considered larger angles of incidence. On the leeward side c_f develops a maximum and a minimum for moderate but not too large values of α. As α increases, the peak and dip in the skin-friction coefficient on the leeward side near the nose becomes more pronounced; the local skin friction vanishes at approximately 42 degrees, indicating separation. This result is in excellent agreement with that computed for the zero-thickness case as described by Cebeci et al. (1980a).

8.17 Results away from the line of symmetry: laminar flows

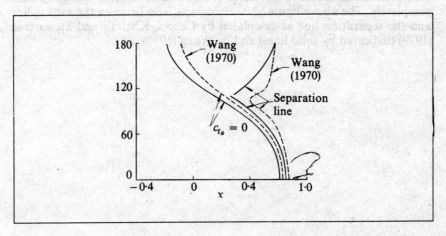

Recently the procedure described in the previous panels has been used to obtain solutions for the general case for both laminar and turbulent flows at $\alpha = 6°$ (see Cebeci, Khattab and Stewartson (1979, 1980b). The governing equations were first solved with the standard Box method until the circumferential velocity component W changed sign, at a position which in the laminar case varied from $x = -0.225$ at $\theta = 180°$ through $x = 0.394$ at $\theta = 90°$ to $x = 0.761$ at $\theta = 0°$. The solution procedure was then switched from the standard Box to the zig-zag Box or to the characteristic Box. Comparison between these two methods has shown that both are satisfactory, but the zig-zag Box is less sensitive to the structural properties of the boundary layer and our final results were, therefore, obtained with the characteristic Box.

In the laminar study the numerical solution begins to develop large gradients in x and θ around $\theta = 115°$ as x approaches 0.30, and at $x = 0.31$ it separates at about this angle, the calculations being started from the windward line of symmetry up to this point. Continuing the calculations in this way, a separation line $\theta = \theta_s(x)$ is found on which θ decreases monotonically until it vanishes at $x = 0.840$. Careful examination of the solution shows, reasonably convincingly, that the separation line is an envelope of limiting streamlines (see Brown, 1965). The calculations were repeated starting from the leeward line of symmetry, a procedure normally accurate but with less margin of safety near separation since the mainstream velocity has a positive component in the direction of θ. For $x \leqslant 0.30$ both methods agree well, but for $0.31 \leqslant x \leqslant 0.34$ the leeward integration can be continued right through to $\theta = 0$ even though separation has occurred in the windward integration. For $x \geqslant 0.35$, the leeward integration breaks down also, at values of θ varying from $110°$ at $x = 0.35$ to $180°$ at $x = 0.72$. The manner of breakdown is, we believe, also associated with a Brown-type singularity but the final breakdown occurs so fast that we were unable to verify this conjecture completely. The above figure shows a comparison between the zero c_f-line and the separation line as calculated by Cebeci, Khattab and Stewartson (1979) (indicated by solid lines) and by Wang (1970).

8.18 Results away from the line of symmetry: laminar flows

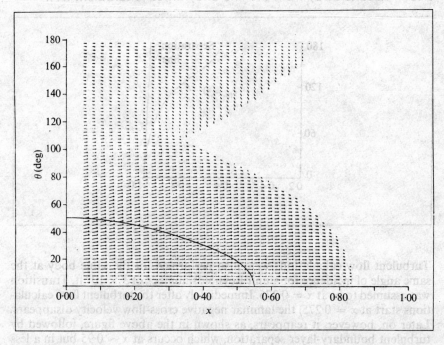

The above figure shows the local skin friction near the separation line by arrows of appropriate length and direction. They bring out very clearly the way these lines are turning against the direction of the main-stream velocity. As can been seen, these results differ significantly from those of Wang; for example, his separation only extends upstream to $x \sim 0.50$, whereas our calculations indicate that separation is initiated near $x = 0.31$.

The arrows in the diagram generally give the magnitude and direction of the skin friction vectors over the aft portion of the spheroid and upstream of the separation line. Below the solid curve, however, the magnitudes of these vectors are so large that for clarity the length of the arrows has been kept constant. In the neighbourhood of $x = 0.31$, the skin friction vectors computed from the windward line of symmetry, where available, have been allowed to override those computed from the leeward line of symmetry.

8.19 Results away from the line of symmetry: turbulent flow

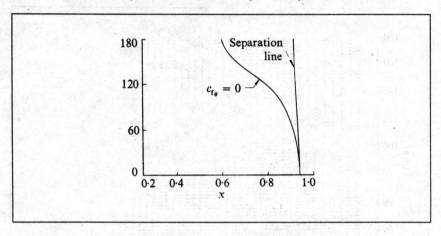

Turbulent flow calculations were also performed for the same body at the same angle of incidence for a unit Reynolds number of 3×10^6 ft. Transition was assumed to be at $x = 0.275$. Immediately after the turbulent flow calculations start at $x = 0.275$) the laminar negative cross-flow velocity disappears. Later on, however, it reappears, as shown in the above figure, followed by turbulent boundary-layer separation, which occurs at $x \sim 0.95$ but in a less spectacular manner than for laminar flow. In the present context turbulent separation is defined by the breakdown of the calculations. It is preceded by a precipitous fall in the streamwise skin friction. In view of the eddy-viscosity model used, the precise manner of breakdown is expected to take the form of a Brown singularity, as in laminar flow. Note that in turbulent flow separation, the laminar "arrow" seen in panel 8.18 is almost absent.

9 Computer Program for Unsteady Two-dimensional Boundary Layers

9.1 Purposes and outline of chapter

- To present a more detailed description of the Box method for solving the two-dimensional boundary-layer equations with and without flow reversal for both laminar and turbulent flows in which the external velocity is a function of both x and t
- To describe a computer program which implements this procedure and uses an eddy-viscosity formulation to model the Reynolds shear stress term
- To demonstrate how the computer program can be easily modified to compute steady inverse flows by using the Mechul-function method with FLARE approximation
- To present calculations for laminar and turbulent flows

In part of Chapter 8 we presented a very brief description of the Box method for solving unsteady boundary-layer equations for two-dimensional flows and as an example we discussed the laminar flow over an impulsively started circular cylinder. Here we present a more detailed explanation of this subject and describe a computer program developed by Cebeci and Carr (1980) for calculating laminar and turbulent boundary layers for an external flow in which the velocity is a function of both x and t. We use the Cebeci–Smith eddy viscosity formulation to model the Reynolds shear-stress term. After describing the main features and the numerical formulation of the method, instructions are provided for the computer program and for each subroutine. We also present sample calculations to demonstrate its use and capabilities for laminar and turbulent flows in which there may be a flow reversal across the boundary layer.

The unsteady computer program described in this chapter can also be used to compute steady, two-dimensional, inverse flows. For this reason, in panels 9.14 through 9.18, we consider the calculation of inverse boundary-layer flows by the Mechul function method with the FLARE approximation discussed in Chapter 7. We show how the computer code for unsteady flows can be modified for this purpose.

9.2 A similarity transformation for unsteady flows

With a transformation similar to that of Falkner–Skan

$$\eta = \sqrt{\frac{\overline{U}_0}{X}}\, Y, \qquad \psi = (X\overline{U}_0)^{1/2} f(X, \eta, T),$$

the nondimensional boundary-layer equations and their boundary conditions (equations (4) to (6) below) can be written as

$$(bf'')' + P_1 ff'' - P_2(f')^2 + P_3 = X\left(f'\frac{\partial f'}{\partial X} - f''\frac{\partial f}{\partial X} + \frac{1}{\overline{U}_0}\frac{\partial f'}{\partial T}\right)$$

$$\eta = 0, \quad f = f' = 0; \qquad \eta = \eta_e, \quad f' = \overline{U}_e/\overline{U}_0$$

Here primes denote differentiation with respect to η and

$$\overline{U}_0(X) = \frac{U_0(X)}{U_\infty}, \quad P_2 = \frac{X}{\overline{U}_0}\frac{d\overline{U}_0}{dX}, \quad P_3 = \frac{X}{\overline{U}_0^2}\left(\overline{U}_e\frac{\partial \overline{U}_e}{\partial X} + \frac{\partial \overline{U}_e}{\partial T}\right)$$

$$P_1 = (P_2 + 1)/2$$

The two-dimensional unsteady boundary-layer equations with an eddy viscosity are

$$\frac{\partial U}{\partial x} + \frac{\partial V}{\partial y} = 0, \tag{1}$$

$$\frac{\partial U}{\partial t} + U\frac{\partial U}{\partial x} + V\frac{\partial U}{\partial y} = \frac{\partial U_e}{\partial t} + U_e\frac{\partial U_e}{\partial x} + \frac{\partial}{\partial y}\left[(\nu + \nu_t)\frac{\partial U}{\partial y}\right]. \tag{2}$$

In terms of dimensionless variables defined by

$$\overline{U} = \frac{U}{U_\infty}, \quad \overline{V} = \frac{V}{U_\infty}\sqrt{R_L}, \quad \overline{U}_e = \frac{U_e}{U_\infty}, \quad Y = \sqrt{\frac{U_\infty}{\nu L}}\, y, \quad X = \frac{x}{L},$$

$$T = \frac{tU_\infty}{L}, \quad b = 1 + \frac{\nu_t}{\nu}, \tag{3}$$

where U_∞ and L denote a reference velocity and length, respectively, equations (1) and (2) and their boundary conditions can be written as

$$\frac{\partial \overline{U}}{\partial X} + \frac{\partial \overline{V}}{\partial Y} = 0 \tag{4}$$

$$\frac{\partial \overline{U}}{\partial T} + \overline{U}\frac{\partial \overline{U}}{\partial X} + \overline{V}\frac{\partial \overline{U}}{\partial Y} = \frac{\partial \overline{U}_e}{\partial T} + \overline{U}_e\frac{\partial \overline{U}_e}{\partial X} + \frac{\partial}{\partial Y}\left[b\frac{\partial \overline{U}}{\partial Y}\right] \tag{5}$$

$$Y = 0, \quad \overline{U} = \overline{V} = 0; \quad Y = Y_e, \quad \overline{U} = \overline{U}_e(X, T). \tag{6}$$

With the introduction of the transformation defined by the above panel, equations (4) to (6) can be written in the form shown on the panel.

9.3 Initial conditions

> - In the (T, η) plane for $X = X_0$ and $T > 0$, solve
> $$(bf'')' + P_1 ff'' - P_2(f')^2 + P_3 = \frac{X}{\bar{U}_0}\frac{\partial f'}{\partial T}$$
> subject to
> $$\eta = 0, \quad f = f' = 0; \quad \eta = \eta_e, \quad f' = \bar{U}_e(X_0, T)$$
> - In the (X, η) plane for $T = 0$ and $X \geq 0$, solve
> $$(bf'')' + P_1 ff'' - P_2(f')^2 + P_3 = X\left(f'\frac{\partial f'}{\partial X} - f''\frac{\partial f}{\partial X}\right)$$
> subject to
> $$\eta = 0, \quad f = f' = 0; \quad \eta = \eta_e, \quad f' = \bar{U}_e(X, 0) \equiv \bar{U}_0$$

To complete the formulation of the problem, initial conditions must be specified in the (T, η) plane for $X = X_0$ and in the (X, η) plane for $T = 0$. Here we consider a flow in which at time $T = 0$ the flow field is given by steady-state conditions and for $T > 0$ the nondimensional external velocity is a function of X and T: this is a very special case, chosen as a simple example.

The initial conditions in the (T, η) and (X, η) planes can be generated by writing the equation of motion and its boundary conditions, as shown above, at $T = 0$.

Note that for a laminar flat-plate flow, the right-hand side of the first equation above is zero; for stagnation-point flow $(\bar{U}_0 = AX)$ it is not. For this reason the slope $A (\equiv d\bar{U}_0/dX)$ must be specified for flows which start as a stagnation point. For details, see Cebeci and Carr (1978). To simplify the discussion and the computer program to be described later, we shall assume that at $X = X_0$ the flow is a flat-plate flow and that for $T > 0$ the solution at $X = X_0$ is the same as the solution at $T = 0$.

If the flow is turbulent at $X = X_0$, $T = 0$, it is necessary to generate initial velocity profiles. This can easily be done by the procedure discussed in panel 5.17. Here, again to simplify the description of the computer program, we shall assume that the flow is laminar at $X = X_0$ for $T = 0$; it can become turbulent at $X > X_0$.

9.4 Description of the computer program, 1: Subroutine ICONZ

> The computer program contains eight subroutines called ICONZ, COEFG, EDDY, GRID, INPUT, OUTPUT, SOLV3, GROWTH and MAIN. The function of each subroutine and MAIN is described here and in the following panels
>
> Subroutine ICONZ contains the finite-difference coefficients of the momentum equation of panel 9.3, for $T = 0$
>
> The linearized difference equations and the boundary conditions are given by equations (4), (5) and (6), below
>
> The calculations are done by marching in the Z-direction for each given X (time)
>
> The Fortran names for the symbols are rather obvious except for G_j, which is designated by V_j. Further, typical correspondences are given below

Symbol	Fortran name
f_j^i	F(J, NZ, 2)
f_j^{i-1}	F(J, NZ $-$ 1, 2)
$f_{j-\frac{1}{2}}^i$	FB
$f_{j-\frac{1}{2}}^{i-1}$	CFB
α_i	BEL
T, X	X, Z

Subroutine ICONZ contains the coefficients of the finite difference equations of the momentum equation for $T = 0$ in panel 9.3. The procedure used to obtain these coefficients is quite straightforward. By introducing the new dependent variables

$$f' = U, \qquad (1a)$$

$$U' = G, \qquad (1b)$$

we write the momentum equation as

$$(bG)' + P_1 fG - P_2 U^2 + P_3 = X\left[\frac{\partial}{\partial X}(U^2/2) - G\frac{\partial f}{\partial X}\right]. \qquad (1c)$$

The finite-difference approximations to equations (1) are

$$f_j^i - f_{j-1}^i - \frac{h_{j-1}}{2}(U_j^i + U_{j-1}^i) = 0, \qquad (2a)$$

$$U_j^i - U_{j-1}^i - \frac{h_{j-1}}{2}(G_j^i + G_{j-1}^i) = 0, \qquad (2b)$$

$$\frac{(bG)^i_j - (bG)^i_{j-1}}{h_{j-1}} + (P_1 + \alpha_i)(fG)^i_{j-\frac{1}{2}} - (P_2 + \alpha_i)(U^2)^i_{j-\frac{1}{2}}$$
$$+ \alpha_i(G^{i-1}_{j-\frac{1}{2}}f^i_{j-\frac{1}{2}} - f^{i-1}_{j-\frac{1}{2}}G^i_{j-\frac{1}{2}}) = R^{i-1}_{j-\frac{1}{2}}. \quad (2c)$$

Here
$$\alpha_i = \frac{X_{i-\frac{1}{2}}}{k_i}, \quad R^{i-1}_{j-\frac{1}{2}} = -L^{i-1}_{j-\frac{1}{2}} - P^i_3 + \alpha_i[(fG)^{i-1}_{j-\frac{1}{2}} - (U^2)^{i-1}_{j-\frac{1}{2}}],$$
$$L^{i-1}_{j-\frac{1}{2}} = [(bG)'_{j-\frac{1}{2}} + P_1(fG)_{j-\frac{1}{2}} - P_2(U^2)_{j-\frac{1}{2}} + P_3]^{i-1}. \quad (3)$$

The linear system for equations (2) can be written and arranged as ($1 \leq j \leq J$)

$$\delta f_j - \delta f_{j-1} - \frac{h_{j-1}}{2}(\delta U_j + \delta U_{j-1}) = (r_1)_j, \quad (4a)$$

$$\delta U_j - \delta U_{j-1} - \frac{h_{j-1}}{2}(\delta G_j + \delta G_{j-1}) = (r_3)_{j-1}, \quad (4b)$$

$$(S_1)_j \delta G_j + (S_2)_j \delta G_{j-1} + (S_3)_j \delta f_j + (S_4)_j \delta f_{j-1} + (S_5)_j \delta U_j + (S_6)_j \delta U_{j-1} = (r_2)_j. \quad (4c)$$

Here, for simplicity, we have dropped the superscript i and defined $(S_k)_j$ and r_k by

$$(S_1)_j = \frac{b_j}{h_{j-1}} + \frac{P_1 + \alpha_i}{2}f_j - \frac{\alpha_i}{2}f^{i-1}_{j-\frac{1}{2}},$$

$$(S_2)_j = -\frac{b_{j-1}}{h_{j-1}} + \frac{P_1 + \alpha_i}{2}f_{j-1} - \frac{\alpha_i}{2}f^{i-1}_{j-\frac{1}{2}},$$

$$(S_3)_j = \left(\frac{P_1 + \alpha_i}{2}\right)G_j + \frac{\alpha_i}{2}G^{i-1}_{j-\frac{1}{2}},$$

$$(S_4)_j = \left(\frac{P_1 + \alpha_i}{2}\right)G_{j-1} + \frac{\alpha_i}{2}g^{i-1}_{j-\frac{1}{2}},$$

$$(S_5)_j = -(P_2 + \alpha_i)U_j,$$

$$(S_6)_j = -(P_2 + \alpha_i)U_{j-1},$$

$$(r_1)_j = f_{j-1} - f_j + h_{j-1}U_{j-\frac{1}{2}}, \quad (5a)$$

$$(r_2)_j = R^{i-1}_{j-\frac{1}{2}} - [(bG)_{j-\frac{1}{2}} + (P_1 + \alpha_i)(fG)_{j-\frac{1}{2}} - (P_2 + \alpha_i)(U^2)_{j-\frac{1}{2}}$$
$$+ \alpha_i(G^{i-1}_{j-\frac{1}{2}}f_{j-\frac{1}{2}} - f^{i-1}_{j-\frac{1}{2}}G_{j-\frac{1}{2}})], \quad (5b)$$

$$(r_3)_{j-1} = U_{j-1} - U_j + h_{j-1}G_{j-\frac{1}{2}}. \quad (5c)$$

The two-wall boundary conditions and the edge boundary condition can be written as

$$\delta f_0 = (r_1)_0 = 0, \quad (6a)$$
$$\delta U_0 = (r_2)_0 = 0, \quad (6b)$$
$$\delta U_J = (r_3)_J = 0. \quad (6c)$$

A listing of this subroutine is given below.

```
      SUBROUTINE ICONZ
      COMMON/BLCO/ NXT,NZT,NX,NZ,NP,NTR,NPT,RL,ETAE,VGP,
     1             A(61),ETA(61),DETA(61)
      COMMON/BLC1/ X(41),Z(61),U0(61),P1(61),P2(61),P3(41,61),UE(41,61)
      COMMON/BLCP/ DELV(61),F(61,61,2),U(61,61,2),V(61,61,2),B(61,61,2)
      COMMON/BLCC/ S1(61),S2(61),S3(61),S4(61),S5(61),S6(61),
     1             R1(61),R2(61),R3(61)
C - - - - - - - - - - - - - - - - - - - - - - - - - - - - - - - - -
      BEL = 0.0
      IF(NZ .GT. 1) BEL = 0.5*(Z(NZ)+Z(NZ-1))/(Z(NZ)-Z(NZ-1))
      P1P   = P1(NZ)+BEL
      P2P   = P2(NZ)+BEL
      DO 30 J=2,NP
C     DEFINITION OF AVERAGED QUANTITIES
      FB    = 0.5*(F(J,NZ,2)+F(J-1,NZ,2))
      UB    = 0.5*(U(J,NZ,2)+U(J-1,NZ,2))
      VB    = 0.5*(V(J,NZ,2)+V(J-1,NZ,2))
      FVB   = 0.5*(F(J,NZ,2)*V(J,NZ,2)+F(J-1,NZ,2)*V(J-1,NZ,2))
      USB   = 0.5*(U(J,NZ,2)**2+U(J-1,NZ,2)**2)
      DERBV = (B(J,NZ,2)*V(J,NZ,2)-B(J-1,NZ,2)*V(J-1,NZ,2))/DETA(J-1)
      IF(NZ .GT. 1) GO TO 10
      CFB   = 0.0
      CUB   = 0.0
      CVB   = 0.0
      CRB   = -P2(NZ)
      GO TO 20
   10 CFB   = 0.5*(F(J,NZ-1,2)+F(J-1,NZ-1,2))
      CUB   = 0.5*(U(J,NZ-1,2)+U(J-1,NZ-1,2))
      CVB   = 0.5*(V(J,NZ-1,2)+V(J-1,NZ-1,2))
      CFVB  = 0.5*(F(J,NZ-1,2)*V(J,NZ-1,2)+F(J-1,NZ-1,2)*V(J-1,NZ-1,2))
      CUSB  = 0.5*(U(J,NZ-1,2)**2+U(J-1,NZ-1,2)**2)
      CDERBV= (B(J,NZ-1,2)*V(J,NZ-1,2)-B(J-1,NZ-1,2)*V(J-1,NZ-1,2))/
     1       DETA(J-1)
      CLB   = CDERBV+P1(NZ-1)*CFVB-P2(NZ-1)*CUSB+P3(NX,NZ-1)
      CRB   = -P3(NX,NZ)+BEL*(CFVB-CUSB)-CLB
C
C     COEFFICIENTS OF THE DIFFERENCED MOMENTUM EQUATION
   20 S1(J) = B(J,NZ,2)/DETA(J-1)+(P1P*F(J,NZ,2)-BEL*CFB)*0.5
      S2(J) = -B(J-1,NZ,2)/DETA(J-1)+(P1P*F(J-1,NZ,2)-BEL*CFB)*0.5
      S3(J) = 0.5*(P1P*V(J,NZ,2)+BEL*CVB)
      S4(J) = 0.5*(P1P*V(J-1,NZ,2)+BEL*CVB)
      S5(J) = -P2P*U(J,NZ,2)
      S6(J) = -P2P*U(J-1,NZ,2)
C
C     DEFINITIONS OF RJ
      R1(J) = F(J-1,NZ,2)-F(J,NZ,2)+DETA(J-1)*UB
      R3(J-1)=U(J-1,NZ,2)-U(J,NZ,2)+DETA(J-1)*VB
      R2(J) = CRB-(DERBV+P1P*FVB-P2P*USB-BEL*(CFB*VB-CVB*FB))
   30 CONTINUE
      R1(1) = 0.0
      R2(1) = 0.0
      R3(NP)= 0.0
      RETURN
      END
```

9.5 Description of the computer program, 2: Subroutine COEFG

> This subroutine contains the finite difference coefficients of the momentum equation and its boundary conditions as given in panel 9.2. We again introduce new dependent variables as given by equations (1a), (1b) and write the second equation of panel 9.2 as
>
> $$(bG)' + P_1 fG - P_2 U^2 + P_3 = X\left[\frac{\partial}{\partial X}\left(\frac{U^2}{2}\right) - G\frac{\partial f}{\partial X} + \frac{1}{U_0}\frac{\partial U}{\partial T}\right] \quad (7)$$
>
> - The finite-difference coefficients for the regular box are given by equations (8) and (9) and for the zig-zag box by equations (10) and (11).
> - Essentially this subroutine is very similar to subroutine ICONZ. The main difference occurs in the coefficients of the finite-difference momentum equation as written in the form given by equation (4c), which now applies to $T > 0$.

Following the procedure discussed in Chapter 8, we write the finite-difference approximations to equation (7) and linearize the resulting non-linear algebraic equations. This allows us to express it in the same form as equation (4c), except now $1 \leqslant j \leqslant J$:

$$(S_1)_j = \frac{b_j}{h_{j-1}} + \frac{\bar{P}_1}{2}f_j + \frac{\alpha_i}{4}(f_{j-\frac{1}{2}} + m_4), \quad (8a)$$

$$(S_2)_j = -\frac{b_{j-1}}{h_{j-1}} + \frac{\bar{P}_1}{2}f_{j-1} + \frac{\alpha_i}{4}(f_{j-\frac{1}{2}} + m_4), \quad (8b)$$

$$(S_3)_j = \frac{\bar{P}_1}{2}G_j + \frac{\alpha_i}{4}(G_{j-\frac{1}{2}} + m_3), \quad (8c)$$

$$(S_4)_j = \frac{\bar{P}_1}{2}G_{j-1} + \frac{\alpha_i}{4}(G_{j-\frac{1}{2}} + m_3), \quad (8d)$$

$$(S_5)_j = -(\bar{P}_2 + \alpha_i)U_j - \beta_k, \quad (8e)$$

$$(S_6)_j = -(\bar{P}_2 + \alpha_i)U_{j-1} - \beta_k, \quad (8f)$$

$$(r_2)_j = (n_3)_j - \{(bG)'_{j-\frac{1}{2}} + \bar{P}_1(fG)_{j-\frac{1}{2}} - (\bar{P}_2 + \alpha_i)(U^2)_{j-\frac{1}{2}}$$

$$+ \frac{\alpha_i}{2}[G_{j-\frac{1}{2}}f_{j-\frac{1}{2}} + m_3 f_{j-\frac{1}{2}} + m_4 G_{j-\frac{1}{2}}] - 2\beta_k U_{j-\frac{1}{2}}\}, \quad (9)$$

where

$$(n_3)_j = m_1\alpha_i - \frac{\alpha_i}{2}m_3 m_4 + 2\beta_k m_2 - m_8 - \bar{P}_1 n_1 + \bar{P}_2 n_2 - 4\bar{P}_3,$$

$$\alpha_i = \frac{X_{i-\frac{1}{2}}}{X_i - X_{i-1}}, \qquad \beta_k = \frac{X_{i-\frac{1}{2}}}{(\overline{U}_0)^{i-\frac{1}{2}}(t_k - t_{k-1})}$$

$$m_1 = (U^2)^{(4)}_{j-\frac{1}{2}} - 2(\overline{U^2})_{i-1}, \qquad m_2 = U^{(2)}_{j-\frac{1}{2}} - 2\overline{U}_{k-1}, \qquad m_3 = G^{234}_{j-\frac{1}{2}},$$

$$m_4 = f^{(4)}_{j-\frac{1}{2}} - 2\bar{f}_{i-1}, \qquad m_8 = [(bG)^{234}_j - (bG)^{234}_{j-1}]/h_{j-1},$$

$$n_1 = (fG)^{234}_{j-\frac{1}{2}}, \qquad n_2 = (U^2)^{234}_{j-\frac{1}{2}},$$

$$\bar{P}_1 = (P_1)^{i-\frac{1}{2}}, \qquad \bar{P}_2 = (P_2)^{i-\frac{1}{2}}, \qquad \bar{P}_3 = (P_3)^{i-\frac{1}{2}}_{k-\frac{1}{2}}.$$

In the region across the boundary layer where $U_j \leq 0$, we replace the regular box method by the zig-zag box discussed in Chapter 8. The only change occurs in the coefficients of equation (4c) of panel 9.4. This time they are defined by

$$(S_1)_j = b_j/h_{j-1} + \frac{\bar{P}_1}{2}f_j + \frac{e_1}{2}(f_{j-\frac{1}{2}} - f^{i-1,k}_{j-\frac{1}{2}}), \tag{10a}$$

$$(S_2)_j = -b_{j-1}/h_{j-1} + \frac{\bar{P}_1}{2}f_{j-1} + \frac{e_1}{2}(f_{j-\frac{1}{2}} - f^{i-1,k}_{j-\frac{1}{2}}), \tag{10b}$$

$$(S_3)_j = \bar{P}_1/2G_j + \frac{e_1}{2}(G_{j-\frac{1}{2}} + G^{i-1,k}_{j-\frac{1}{2}}), \tag{10c}$$

$$(S_4)_j = \bar{P}_1/2G_{j-1} + \frac{e_1}{2}(G_{j-\frac{1}{2}} + G^{i-1,k}_{j-\frac{1}{2}}), \tag{10d}$$

$$(S_5)_j = -(\bar{P}_2 + e_1)U_j - e_2, \tag{10e}$$

$$(S_6)_j = -(\bar{P}_2 + e_1)U_{j-1} - e_2, \tag{10f}$$

$$(r_2)_j = C_3 - [(bG)_{j-\frac{1}{2}} + \bar{P}_1(fG)_{j-\frac{1}{2}} - \bar{P}_2(U^2)_{j-\frac{1}{2}} - 2e_2 U_{j-\frac{1}{2}} - e_1[(U^2)_{j-\frac{1}{2}}$$
$$- (G_{j-\frac{1}{2}}f_{j-\frac{1}{2}} + G^{i-1,k}_{j-\frac{1}{2}}f_{j-\frac{1}{2}} - f^{i-1,k}_{j-\frac{1}{2}}G_{j-\frac{1}{2}})], \tag{11}$$

$$C_3 = -C_2 + 2C_1 - 2e_2 U^{i,k-1}_{j-\frac{1}{2}},$$

$$C_1 = -e_1[\tfrac{1}{2}G^{i-1,k}_{j-\frac{1}{2}}f^{i-1,k}_{j-\frac{1}{2}} - \tfrac{1}{2}(U^2)^{i-1,k}_{j-\frac{1}{2}}] + e_3[\tfrac{1}{2}(U^2)^{i,k-1}_{j-\frac{1}{2}} - \tfrac{1}{2}(U^2)^{i+1,k-1}_{j-\frac{1}{2}}$$
$$- G^{i+\frac{1}{2},k-1}_{j-\frac{1}{2}}(f^{i,k-1}_{j-\frac{1}{2}} - f^{i+1,k-1}_{j-\frac{1}{2}})],$$

$$C_2 = \frac{(bG)^{i,k-1}_j - (bG)^{i,k-1}_{j-1}}{h_{j-1}} + \bar{P}_1(fG)^{i,k-1}_{j-\frac{1}{2}} - \bar{P}_2(U^2)^{i,k-1}_{j-\frac{1}{2}} + 2\bar{P}_3,$$

$$e_1 = \frac{X_i \theta}{X_i - X_{i-1}}, \qquad e_2 = \frac{X_i}{(\overline{U}_0)^{i-\frac{1}{2}}(t_k - t_{k-1})}, \qquad e_3 = \frac{X_i \phi}{X_i - X_{i+1}}.$$

Computer Program for Unsteady Two-dimensional Boundary Layers

A listing of this subroutine is given below.

```
      SUBROUTINE COEFG
      COMMON/BLC0/ NXT,NZT,NX,NZ,NP,NTR,NPT,RL,ETAE,VGP,
     1             A(61),ETA(61),DETA(61)
      COMMON/BLC1/ X(41),Z(61),UO(61),P1(61),P2(61),P3(41,61),UE(41,61)
      COMMON/BLCP/ DELV(61),F(61,61,2),U(61,61,2),V(61,61,2),B(61,61,2)
      COMMON/BLCG/ S1(61),S2(61),S3(61),S4(61),S5(61),S6(61),
     1             R1(61),R2(61),R3(61)
      COMMON/ZGZG/ KALC(61)
C - - - - - - - - - - - - - - - - - - - - - - - - - - - - - - - - - -
      U(NP,NZ,2) = UE(NX,NZ) / UO(NZ)
      UOB      = 0.5*(UO(NZ)+UO(NZ-1))
      KALC(1)=0
      DELX     = X(NX)-X(NX-1)
      DELZ     = Z(NZ)-Z(NZ-1)
      ZB       = 0.5*(Z(NZ)+Z(NZ-1))
      CEL      = ZB/DELZ
      BELM     = ZB/(DELX)/UOB
      CEL2     = 0.5*CEL
      CEL4     = 0.25*CEL
      P1H      = 0.5*P1(NZ)
      P2H      = 0.5*P2(NZ)
      P1B      = 0.5*(P1(NZ)+P1(NZ-1))
      P2B      = 0.5*(P2(NZ)+P2(NZ-1))
      P1BH     = 0.5*P1B
      P3B      = 0.25*(P3(NX,NZ)+P3(NX-1,NZ)+P3(NX,NZ-1)+P3(NX-1,NZ-1))
      IF(NZ .EQ. NZT) GO TO 10
      NZP1     = NZ+1
      E1       = Z(NZ)*( (Z(NZP1)-Z(NZ))/(Z(NZP1)-Z(NZ-1))/DELZ )
      E1H      = 0.5*E1
      E2       = Z(NZ)/DELX/UO(NZ)
      E3       = Z(NZ)*DELZ/(Z(NZP1)-Z(NZ-1))/(Z(NZ)-Z(NZP1))
      P3BZ     = 0.5*( P3(NX,NZ)+P3(NX-1,NZ) )
   10 CONTINUE
      DO 50  J=2,NP
      FB       = 0.5*(F(J,NZ,2)+F(J-1,NZ,2))
      FVB      = 0.5*(F(J,NZ,2)*V(J,NZ,2)+F(J-1,NZ,2)*V(J-1,NZ,2))
      UB       = 0.5*(U(J,NZ,2)+U(J-1,NZ,2))
      USB      = 0.5*(U(J,NZ,2)**2+U(J-1,NZ,2)**2)
      UB2      = 0.5*(U(J,NZ-1,2)+U(J-1,NZ-1,2))
      UB4      = 0.5*(U(J,NZ,1)+U(J-1,NZ,1))
      VB       = 0.5*(V(J,NZ,2)+V(J-1,NZ,2))
      DERBV    = (B(J,NZ,2)*V(J,NZ,2)-B(J-1,NZ,2)*V(J-1,NZ,2))/DETA(J-1)
      FB4      = 0.5*(F(J,NZ,1)+F(J-1,NZ,1))
      USB4     = 0.5*(U(J,NZ,1)**2+U(J-1,NZ,1)**2)
      IF(NZ .EQ. NZT) GO TO 20
      IF ( UB .LT. 0.0 ) GO TO 30
   20 FVJ2     = F(J,NZ,1)*V(J,NZ,1)+F(J,NZ-1,1)*V(J,NZ-1,1)+
     1           F(J,NZ-1,2)*V(J,NZ-1,2)
      FVJ1     = F(J-1,NZ,1)*V(J-1,NZ,1)+F(J-1,NZ-1,1)*V(J-1,NZ-1,1)+
     1           F(J-1,NZ-1,2)*V(J-1,NZ-1,2)
      FBI1     = 0.25*(F(J,NZ-1,1)+F(J-1,NZ-1,1)+F(J,NZ-1,2)+F(J-1,NZ-1,2))
      FVB234   = 0.5*(FVJ2+FVJ1)
      USB2     = 0.5*(U(J,NZ-1,2)**2+U(J-1,NZ-1,2)**2)
      USB3     = 0.5*(U(J,NZ-1,1)**2+U(J-1,NZ-1,1)**2)
      USBI1    = 0.5*(USB2+USB3)
      UBK1     = 0.25*(U(J,NZ-1,1)+U(J-1,NZ-1,1)+U(J,NZ,1)+U(J-1,NZ,1))
      USJ2     = U(J,NZ,1)**2+U(J,NZ-1,1)**2+U(J,NZ-1,2)**2
```

```
            USJ1   = U(J-1,NZ,1)**2+U(J-1,NZ-1,1)**2+U(J-1,NZ-1,2)**2
            USB234= 0.5*(USJ2+USJ1)
            VJ1    = V(J-1,NZ,1)+V(J-1,NZ-1,1)+V(J-1,NZ-1,2)
            VJ2    = V(J,NZ,1)+V(J,NZ-1,1)+V(J,NZ-1,2)
            VB234  = 0.5*(VJ2+VJ1)
            BVJ1   = B(J-1,NZ,1)*V(J-1,NZ,1)+B(J-1,NZ-1,1)*V(J-1,NZ-1,1)+
           1         B(J-1,NZ-1,2)*V(J-1,NZ-1,2)
            BVJ2   = B(J,NZ,1)*V(J,NZ,1)+B(J,NZ-1,1)*V(J,NZ-1,1)+
           1         B(J,NZ-1,2)*V(J,NZ-1,2)
      C
            CM1    = USB4-2.0*USBI1
            CM2    = UB2-2.0*UBK1
            CM4    = FB4-2.0*FBI1
            CM8    = (BVJ2-BVJ1)/DETA(J-1)
            CN3J   = CM1*CEL-0.5*CEL*VB234*CM4+2.0*BELM*CM2-CM8-P1B*FVB234
           1         + P2B*USB234-4.0*P3B
      C     COEFFICIENTS FOR THE REGULAR BOX.
            S1(J)  = B(J,NZ,2)/DETA(J-1)+P1BH*F(J,NZ,2)+CEL4*(FB+CM4)
            S2(J)  =-B(J-1,NZ,2)/DETA(J-1)+P1BH*F(J-1,NZ,2)+CEL4*(FB+CM4)
            S3(J)  = P1BH*V(J,NZ,2)+CEL4*(VB+VB234)
            S4(J)  = P1BH*V(J-1,NZ,2)+CEL4*(VB+VB234)
            S5(J)  =-(P2B+CEL)*U(J,NZ,2)-BELM
            S6(J)  =-(P2B+CEL)*U(J-1,NZ,2)-BELM
            R2(J)  = CN3J-(DERBV+P1B*FVB-(P2B+CEL)*USB+CEL2*(VB*FB+VB234*FB
           1         +CM4*VB)-BELM*2.0*UB)
            KALC(J)=0
            GO TO 40
      C
         30 FB2    = 0.5*(F(J,NZ-1,2)+F(J-1,NZ-1,2))
            VB2    = 0.5*(V(J,NZ-1,2)+V(J-1,NZ-1,2))
            FVB4   = 0.5*(F(J,NZ,1)*V(J,NZ,1)+F(J-1,NZ,1)*V(J-1,NZ,1))
            UB6    = 0.5*(U(J,NZP1,1)+U(J-1,NZP1,1))
            FB6    = 0.5*(F(J,NZP1,1)+F(J-1,NZP1,1))
            UB46   = 0.25*(U(J,NZ,1)+U(J-1,NZ,1)+U(J,NZP1,1)+U(J-1,NZP1,1))
            VB46   = 0.25*(V(J,NZ,1)+V(J-1,NZ,1)+V(J,NZP1,1)+V(J-1,NZP1,1))
            DRBV4  = (B(J,NZ,1)*V(J,NZ,1)-B(J-1,NZ,1)*V(J-1,NZ,1))/DETA(J-1)
            C1     = E1H*(VB2*FB2-UB2**2)+E3*((UB4-LB6)*UB46-VB46*(FB4-FB6))
            C2     = DRBV4+P1(NZ)*FVB4-P2(NZ)*USB4+2.0*P3BZ
            C3     = -C2+2.0*C1-2.0*E2*UB4
      C     COEFFICIENTS FOR THE ZIG-ZAG BOX.
            S1(J)  = B(J,NZ,2)/DETA(J-1)+P1H*F(J,NZ,2)+E1H*(FB-FB2)
            S2(J)  = -B(J-1,NZ,2)/DETA(J-1)+P1H*F(J-1,NZ,2)+E1H*(FB-FB2)
            S3(J)  = P1H*V(J,NZ,2)+E1H*(VB+VB2)
            S4(J)  = P1H*V(J-1,NZ,2)+E1H*(VB+VB2)
            S5(J)  = -P2(NZ)*U(J,NZ,2)-E2-E1*U(J,NZ,2)
            S6(J)  = -P2(NZ)*U(J-1,NZ,2)-E2-E1*U(J-1,NZ,2)
            R2(J)  = C3-(DERBV+P1(NZ)*FVB-P2(NZ)*USB-2.0*E2*UB-E1*(UB**2
           1         -VB*FB-VB2*FB+FB2*VB))
            KALC(J)=1
         40 R1(J)  = F(J-1,NZ,2)-F(J,NZ,2)+DETA(J-1)*UB
            R3(J-1)=U(J-1,NZ,2)-U(J,NZ,2)+DETA(J-1)*VB
         50 CONTINUE
            R3(NP) = 0.0
            R1(1)  = 0.0
            P2(1)  = 0.0
            RETURN
            END
```

9.6 Subroutine EDDY

- Inner layer

$$\left(\frac{v_t}{v}\right)_i = 0{\cdot}16\sqrt{R_L}(\bar{U}_0 X)^{1/2}|f''|(1 - e^{-y^+/A^+})^2$$

- Outer layer

$$\left(\frac{v_t}{v}\right)_o = \alpha\sqrt{R_L}(\bar{U}_0 X)^{1/2}(f'_e\eta_e - f_e)$$

where

$$\frac{y^+}{A^+} = \frac{R_L^{1/4}(\bar{U}_0 X)^{1/4}}{26}\eta(f''_w)^{1/2}(1 - 11{\cdot}8\,p_x^+)^{1/2}$$

and

$$p_x^+ = \frac{P_2}{R_L^{1/4}(\bar{U}_0 X)^{1/4}}\left(\frac{\bar{U}_e}{\bar{U}_0}\right)^2\frac{1}{(f''_w)^{3/2}}$$

The formulas for the Cebeci–Smith eddy viscosity formulation are given by panel 3.6. To make this subroutine short and simple, we consider the case of no mass transfer, neglect the intermittency term γ_{tr} and assume α in the outer eddy viscosity formula to be constant ($\equiv 0{\cdot}0168$). In terms of transformed variables, we then write the formulas as shown in the above panel.

The inner and outer regions for the eddy viscosity formulas are established by the continuity of the eddy viscosity: we use $(v_t/v)_i$ for the inner region until it exceeds $(v_t/v)_0$, then we switch to the expression $(v_t/v)_0$.

It should be noted that when there is a flow reversal across the boundary layer $(v_t/v)_i$, which must always increase with increasing η, may start to decrease and may even become zero since $|f''| \to 0$. This unacceptable behaviour can be avoided by checking the magnitude of $(v_t/v)_i$ at each grid point in the inner layer. If $(v_t/v)_i$ at η_j is less than its value at η_{j-1}, then it can be made to increase by linear extrapolation as is done in Subroutine EDDY.

The Fortran names for some of the symbols are:

Symbol	Fortran name
$(v_t/v)_i, (v_t/v)_0$	EDVI, EDVO
$y^+/A^+ (\equiv y/A)$	YOA
p_x^+	PPLUS

A listing of this subroutine is given below.

```
      SUBROUTINE EDDY
      COMMON/BLCO/ NXT,NZT,NX,NZ,NP,NTR,NPT,RL,ETAE,VGP,
     1             A(61),ETA(61),DETA(61)
      COMMON/BLC1/ X(41),Z(61),UO(61),P1(61),P2(61),P3(41,61),UE(41,61)
      COMMON/BLCP/ DELV(61),F(61,61,2),U(61,61,2),V(61,61,2),B(61,61,2)
      DIMENSION    EDV(61)
C - - - - - - - - - - - - - - - - - - - - - - - - - - - - -
   20 IFLG  = 0
      RZ2   = SQRT (UO(NZ)*Z(NZ)*RL)
      RZ4   = SQRT (RZ2)
      VMAX  = V(1,NZ,2)
      DO 30 J = 2,NP
      IF(ABS(V(J,NZ,2)).GT.ABS(VMAX)) VMAX= V(J,NZ,2)
   30 CONTINUE
      EDVO  = 0.0168*RZ2*(U(NP,NZ,2)*ETA(NP)-F(NP,NZ,2))
      J     = 1
   80 IF(IFLG .EQ. 1) GO TO 90
      PPLUS = (P2(NZ)/RZ4)*(UE(NX,NZ)/UO(NZ))**2*(1.0/V(1,NZ,2)**1.5)
      YOA   = RZ4*ETA(J)*SQRT(V(1,NZ,2)*(1.0-11.8*PPLUS))/26.0
      EL    = 1.0
      IF(YOA .LT. 4.0) EL = (1.0-EXP(-YOA))**2
      EDVI  = 0.16*RZ2*ABS(V(J,NZ,2))*EL*ETA(J)**2
      IF(EDVI .LT. EDVO) GO TO 100
      IFLG  = 1
   90 EDV(J)= EDVO
      GO TO 110
  100 EDV(J)= EDVI
      IF(J.LE.2) GOTO 110
      IF(EDV(J).GT.EDV(J-1)) GOTO 110
      EDV(J)= EDV(J-1)+(EDV(J-1)-EDV(J-2))*VGP
      IF(EDV(J).LT.EDVO) GOTO 110
      EDV(J)= EDVO
      IFLG  = 1
  110 B(J,NZ,2)= 1.0+EDV(J)
      J     = J+1
      IF(J .LE. NP) GO TO 80
      RETURN
      END
```

9.7 Subroutine GRID

- This subroutine generates the grid normal to the flow. It can be either uniform or nonuniform. Since the Box scheme uses only two values of y in each finite difference "molecule", we can use a variety of grids across the layer. The nonuniform one generated in this subroutine is described below
- This subroutine also generates the initial guess profiles for the similarity equation in the laminar case, described below. With additional changes one can also generate the initial velocity profiles for turbulent flows

While there are a number of choices for generating a nonuniform grid across the layer, here we use one which is a geometric progression with the property that the ratio of lengths of any two adjacent intervals is a constant; that is, $\Delta \eta_j = K \Delta \eta_{j-1}$. The distance to the jth line is given by the formula:

$$\eta_j = \Delta \eta_1 (K^j - 1)/(K - 1), \qquad K > 1. \tag{1}$$

There are two parameters in the above equation: $\Delta \eta_1$, the length of the first step, and K, the ratio of two successive steps. The total number of points J can be calculated from the following formula:

$$J = \frac{\ln \left[1 + (K - 1)(\eta_e/\Delta \eta_1)\right]}{\ln K}. \tag{2}$$

For moderate Reynolds numbers $\Delta \eta_1$ and K are chosen typically as 0·01 and 1·14, respectively. In general, approximately 50 grid points across the boundary layer are sufficient to represent laminar and turbulent boundary layer flows, so the present computer program restricts the number of points across the boundary layer to 61. Consequently, the chosen values of $\Delta \eta_1 (\equiv h_1)$ and K (names DETA(1) and VGP, respectively, in the code) must be such that the number of points does not exceed 61. Figure 1 is presented, therefore, to provide guidance in the selection of J.

Since we are solving a set of nonlinear equations, it is necessary to provide initial guess profiles for the similarity equation

$$f''' + P_1 f f'' - P_2 (f')^2 + P_3 = 0.$$

In this subroutine the initial guess for the velocity profile is assumed to be given by a polynomial

$$f' = a + b\eta + c\eta^3.$$

172 Engineering Calculation Methods for Turbulent Flow

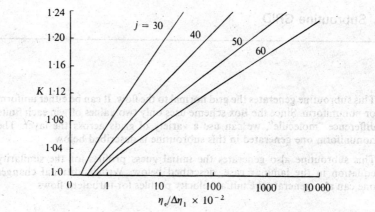

Fig. 1. Variation of K with $\Delta\eta_1$ for different η_e-values.

The constants a, b, and c are chosen so that the boundary conditions $f(0) = 0$, $f'(0) = 0$, $f'(\eta_e) = 1$ are satisfied plus the additional condition $f''(\eta_e) = 0$. Then it follows that

$$f = \frac{\eta_e}{4}\left(\frac{\eta}{\eta_e}\right)^2\left[3 - \tfrac{1}{2}\left(\frac{\eta}{\eta_e}\right)^2\right],$$

$$f' = \tfrac{3}{2}\left(\frac{\eta}{\eta_e}\right) - \tfrac{1}{2}\left(\frac{\eta}{\eta_e}\right)^3,$$

$$f'' = \tfrac{3}{2}\frac{1}{\eta_e}\left[1 - \left(\frac{\eta}{\eta_e}\right)^2\right].$$

A listing of this subroutine is given below.

```
      SUBROUTINE GRID
      COMMON/BLC0/ NXT,NZT,NX,NZ,NP,NTR,NPT,RL,ETAE,VGP,
     1             A(61),ETA(61),DETA(61)
      COMMON/BLC1/ X(41),Z(61),U0(61),P1(61),P2(61),P3(41,61),UE(41,61)
      COMMON/BLCP/ DELV(61),F(61,61,2),U(61,61,2),V(61,61,2),B(61,61,2)
C - - - - - - - - - - - - - - - - - - - - - - - - - - - - - - - -
      IF((VGP-1.0) .LE. 0.001) GO TO 10
      NP       = ALOG((ETAE/DETA(1))*(VGP-1.0)+1.0)/ALOG(VGP) + 1.0001
      DETA(1)  = ETAE*(VGP-1.0)/(VGP**(NP-1)-1.0)
      GO TO 20
   10 NP       = ETAE/DETA(1) + 1.0001
   20 IF(NP .LE. 61) GO TO 30
      WRITE(6, 50 )
      STOP
   30 ETA(1)= 0.0
      DO 40 J=2,61
      DETA(J)=VGP*DETA(J-1)
      A(J)    = 0.5*DETA(J-1)
   40 ETA(J)= ETA(J-1)+DETA(J-1)
      NZ       = 1
      ETANPQ = 0.25 * ETA(NP)
      ETAU15 = 1.5 / ETA(NP)
      DO 45 J = 1, NP
      ETAB   = ETA(J)/ETA(NP)
      ETAB2  = ETAB**2
      F(J,NZ,2) = ETANPQ*ETAB2*(3.0-0.5*ETAB2)
      U(J,NZ,2) = 0.5*ETAB*(3.0-ETAB2)
      V(J,NZ,2) = ETAU15*(1.0-ETAB2)
      B(J,NZ,2) = 1.0
   45 CONTINUE
      RETURN
C - - - - - - - - - - - - - - - - - - - - - - - - - - - - - - - -
   50 FORMAT(1H0,' NP EXCEEDED 61 -- PROGRAM TERMINATED')
      END
```

9.8 Subroutine INPUT

> - In this subroutine we input (or define) the external velocity distribution $\bar{U}_e(X, T)$ and $\bar{U}_0(X)$ and compute the dimensionless pressure-gradient parameters P_1, P_2 and P_3, as described below
> - We also specify the total number of X-stations (NZT), the total number of T-stations (NXT), the location of transition (NTR) and the reference Reynolds number, $R_L(\equiv U_\infty L/\nu)$

Depending on how one wishes to input the external velocity distribution and compute the surface distance, Subroutine INPUT can be arranged in a number of ways. Here, for simplicity, we shall choose the most convenient and simplest way: we shall assume that the input X-values are surface distances. Rather than inputting external velocity in tabular form, we will assume that values are given by formulas. Also, rather than evaluating the derivatives numerically, say by using three-point Lagrange interpolation formulas (see Cebeci and Bradshaw, 1977, page 261), we shall evaluate them analytically.

For our example we shall assume that the external velocity $\bar{U}_e$ is given by the following formula, which was used for laminar and turbulent flows by Cebeci (1978) and Cebeci, Carr and Bradshaw (1979):

$$\bar{U}_e = 1 - \alpha(X - X^2)(T^2 - T^3), \qquad 0 < X < 1. \tag{1}$$

Here α is a positive constant equal to say, 20.

In order to compute the boundary layer due to (1), initial conditions in the (T, η) and (X, η) planes are required and may be obtained by the method of panel (9.3). If $X_0 = 0$, the initial profile can be taken as Blasius and there is no difficulty about computing the solution in $X > 0$ since the initial boundary layer is of zero thickness. If $X_0 \neq 0$ we could take

$$\bar{U}_e = 1 - \alpha(X_0 - X_0^2)(T^2 - T^3), \qquad 0 < X < X_0, \tag{2}$$

but then there is a discontinuity in the pressure gradient at $X = X_0$. Since it acts on an already established boundary layer, the initial response is inviscid leading formally to a velocity slip and hence a subboundary layer at the wall. The treatment of the boundary layer is then rather subtle (Cebeci, Stewartson and Williams, 1979) but if we are not too concerned with the details of the solution, a convenient procedure near $X = X_0$ is to write (1) as

$$\bar{U}_e = 1 - \alpha F[(X - X_0)/a](X - X^2)(T^2 - T^3) \tag{3}$$

where F is a smooth function which vanishes if $X < X_0$ and is unity if $X - X_0 > a$. For example, take $F(s) = \sin \pi s/2, 0 < s < 1$, and $a = 0.06$ with 10 stations between $X = X_0$ and $X = X_0 + a$. A similar difficulty would occur at $T = 0$ if T^2 were replaced by T since the boundary layer is well established at $T = 0$.

In this subroutine we use the expressions given by equations (1) and (2) and compute the dimensionless pressure gradients P_1, P_2 and P_3. In addition we specify NZT, NXT and NTR and read in R_L as shown in the above panel.

A listing of this subroutine is given below.

```
      SUBROUTINE INPUT
      COMMON/BLC0/ NXT,NZT,NX,NZ,NP,NTR,NPT,RL,ETAE,VGP,
     1             A(61),ETA(61),DETA(61)
      COMMON/BLC1/ X(41),Z(61),U0(61),P1(61),P2(61),P3(41,61),UE(41,61)
      COMMON/BLCP/ DELV(61),F(61,61,2),U(61,61,2),V(61,61,2),B(61,61,2)
      DIMENSION    TITLE(20),ZC(61),XC(61)
      DATA         PIH / 1.570796325E0 /,     SA / 0.06E0 /
C - - - - - - - - - - - - - - - - - - - - - - - - - - - - - - - - - -
      NPT      = 61
      ETAE     = 8.0
      DETA(1)  = 0.2
      VGP      = 1.0
      ALPHA    = 20.0
      READ(5, 260) NXT,NZT,NTR,RL
      READ(5, 290 ) (X(I),I=1,NXT)
      READ(5, 290 ) (Z(I),I=1,NZT)
      DO 150 I=1,NZT
      ZM1      = Z(I)-1.0
      ZM1S     = ZM1*ZM1
      T        = PIH*(ZM1/SA)
      SINT     = SIN( T )
      COST     = COS( T )
      DO 150 K=1,NXT
      X2MX3 = X(K)**2-X(K)**3
      IF ( ZM1 .GT. SA ) GO TO 80
      UE(K,I)= 1.0-ALPHA*SINT*(ZM1-ZM1S)*X2MX3
      DUEDX =-ALPHA*X2MX3*( SINT*(1.0-2.0*ZM1)+(ZM1-ZM1S)*(PIH/SA)*COST)
      DUEDT =-ALPHA*SINT*(ZM1-ZM1S)*(2.0*X(K)-3.0*X(K)**2)
      GO TO 100
   80 UE(K,I)= 1.0-ALPHA*(ZM1-ZM1S)*X2MX3
      DUEDX =-ALPHA*X2MX3*( 1.0-2.0*ZM1)
      DUEDT =-ALPHA*(ZM1-ZM1S)*(2.0*X(K)-3.0*X(K)**2)
  100 IF ( K .GT. 1 ) GO TO 120
      U0(I)    = UE(1,I)
      P2(I)    = (Z(I)/U0(I)) * DUEDX
      P1(I)    = 0.5*(1.0+P2(I) )
  120 P3(K,I)= (Z(I)/U0(I)**2) * ( UE(K,I)*DUEDX+DUEDT )
  150 CONTINUE
C
C
      CALL GRID
      RETURN
C - - - - - - - - - - - - - - - - - - - - - - - - - - - - - - - - - -
  260 FORMAT( 3I5,5X,F10.5)
  290 FORMAT(8F10.0)
      END
```

9.9 Subroutine OUTPUT

This subroutine prints out the desired profiles such as f_j, U_j, G_j and b_j as functions of η_j. It also computes the boundary-layer parameters θ, δ^*, H, c_f, R_θ and R_{δ^*}. The integration is made by the trapezoidal rule

The Fortran names for some of the symbols are:

Symbol	Fortran Name
$X/(R_L U_0)^{1/2}$	C_1
$\delta^* = \int_0^\infty \left(1 - \dfrac{U}{U_e}\right) dy = C_1(\eta_e - f_e)/f'_e$	DELSTR
$\theta = \int_0^\infty \dfrac{U}{U_e}\left(1 - \dfrac{U}{U_e}\right) dy = C_1 \int_0^{\eta_e} \dfrac{f'}{f'_e}\left(1 - \dfrac{f'}{f'_e}\right) d\eta$	THETA
$H = \delta^*/\theta$	H
$c_f = \dfrac{2\tau_w}{\rho U_0^2} = \dfrac{2f''_w}{(R_L X U_0)^{1/2}}\left(\dfrac{U_e}{U_0}\right)^2$	CF
$R_\theta = \dfrac{U_e \theta}{\nu}$	RTHETA
$R_{\delta^*} = \dfrac{U_e \delta^*}{\nu}$	RDELST

A listing of this subroutine is given below.

```
      SUBROUTINE OUTPUT
      COMMON/BLC0/ NXT,NZT,NX,NZ,NP,NTR,NPT,RL,ETAE,VGP,
     1             A(61),ETA(61),DETA(61)
      COMMON/BLC1/ X(41),Z(61),U0(61),P1(61),P2(61),P3(41,61),UE(41,61)
      COMMON/BLCP/ DELV(61),F(61,61,2),U(61,61,2),V(61,61,2),B(61,61,2)
      COMMON/ZGZG/ KALC(61)
      DIMENSION RTHETA(61),RTHT(61),NPK(61)
```

```
      WRITE(6, 220 )
      NPM1 = NP-1
      WRITE(6, 230 )(J,ETA(J),F(J,NZ,2),U(J,NZ,2),V(J,NZ,2),B(J,NZ,2),
     1              KALC(J),J=1,NPM1,3)
      J = NP
      WRITE(6, 230 ) J,ETA(J),F(J,NZ,2),U(J,NZ,2),V(J,NZ,2),B(J,NZ,2),
     1              KALC(J)
    5 CONTINUE
      NPK(NZ)=NP
      IF ( NZ .EQ. 1 ) GO TO 140
      C1     = SQRT( Z(NZ) / (RL*UO(NZ)) )
      DELSTR= C1*(ETA(NP)-F(NP,NZ,2)/U(NP,NZ,2))
      CF    =2.0*V(1,NZ,2)/(SQRT(RL*UO(NZ)*Z(NZ))*(UE(NX,NZ)/UC(NZ))**2)
      RDELST= UE(NX,NZ)*DELSTR*RL
      SUM1  = 0.0
      F1    = U(1,NZ,2)/U(NP,NZ,2)*(1.0-U(1,NZ,2)/U(NP,NZ,2))
      DO 10  J=2,NP
      F2    = U(J,NZ,2)/U(NP,NZ,2)*(1.0-U(J,NZ,2)/U(NP,NZ,2))
      SUM1  = SUM1+(F1+F2)*A(J)
   10 F1    = F2
      THETA = C1*SUM1
      RTHETA(NZ) = UE(NX,NZ)*THETA*RL
      H     = DELSTR/THETA
   90 CALL GROWTH(1)
      WRITE ( 6, 242 ) DELSTR,THETA,CF,RDELST,RTHETA(NZ),H,
     1                UE(NX,NZ),P1(NZ),P2(NZ),P3(NX,NZ)
      IF ( NZ .EQ. NZT ) GO TO 190
  140 NZ   = NZ+1
C     INITIAL GUESS FOR NEXT STATION
      DO 150 J=1,NPT
      F(J,NZ,2)= F(J,NZ-1,2)
      U(J,NZ,2)= U(J,NZ-1,2)
      V(J,NZ,2)= V(J,NZ-1,2)
  150 B(J,NZ,2)= B(J,NZ-1,2)
      IF ( NX .EQ. 1 ) RETURN
C
  160 NP     = NPK(NZ)
      IF(NZ.EQ.1) GOTO 170
      IF(NP .LT. NPK(NZ-1)) NP = NPK(NZ-1)
  170 RETURN
  190 IF(NX .EQ. NXT) STOP
      NX    = NX+1
      NZ    = 1
C     SHIFT.
      DO 210 K=1,NZT
      DO 200 J=1,NPT
      F(J,K,1)= .F(J,K,2)
      U(J,K,1)= U(J,K,2)
      V(J,K,1)= V(J,K,2)
  200 B(J,K,1)= B(J,K,2)
  210 CONTINUE
      GO TO 160
C - - - - - - - - - - - - - - - - - - - - - - - - - - - - - -
  220 FORMAT(1H0,2X,1HJ,4X,3HETA,10X,1HF,13X,1HU,13X,1HV,13X,1HB,8X,
     1         4HKALC)
  230 FORMAT(1H ,I3,F10.5,4E14.6,I6)
  242 FORMAT(1H0,7HDELSTR=,E14.6,3X,7HTHETA =,E14.6,3X,7HCF     =,E14.6/
     1           1H ,7HRDELST=,E14.6,3X,7HRTHETA=,E14.6                /
     2           1H ,7HH     =,E14.6,3X,7HUE    =,E14.6,3X,7HP1    =,E14.6/
     3           1H ,7HP2    =,E14.6,3X,7HP3    =,E14.6 / )
      END
```

9.10 Subroutine SOLV3

This subroutine contains the recursion formulas that arise in the block elimination method discussed in panel 5.13. This is a very useful subroutine, since it can be used to solve any third-order ordinary (linear or nonlinear) and/or parabolic partial differential equation as described in Cebeci and Bradshaw (1977)

To solve the linear system given by equations (4) and (5) of panel 9.4 with $(S_k)_j$ and $(r_k)_j$ given by those in panels 9.4 and/or 9.5, we write the linear system as

$$A\delta = r \tag{1}$$

and solve it by the block elimination method discussed in panel 5.13. While the matrices of that panel are 2×2 blocks (since the example problem was a second-order differential equation) the blocks are now 3×3. This introduces no difficulties. As before, we obtain the solution of equation (1) in two sweeps by the general equations given in panel 5.13. Note that the first two rows of the A_0-matrix contain the two wall boundary conditions. If we use boundary conditions other than these two, all we need to do is to alter these two rows of the A_0-matrix.

This subroutine is identical to the SOLV3 given in Cebeci and Bradshaw (1977) where a detailed description is provided.

A listing of this subroutine is given below.

```
      SUBROUTINE SOLV3
      COMMON/BLC0/ NXT,NZT,NX,NZ,NP,NTR,NPT,RL,ETAE,VGP,
     1             A(61),ETA(61),DETA(61)
      COMMON/BLC1/ X(41),Z(61),U0(61),P1(61),P2(61),P3(41,61),UE(41,61)
      COMMON/BLCP/ DELV(61),F(61,61,2),U(61,61,2),V(61,61,2),B(61,61,2)
      COMMON/BLCC/ S1(61),S2(61),S3(61),S4(61),S5(61),S6(61),
     1             R1(61),R2(61),R3(61)
      DIMENSION A11(61),A12(61),A13(61),A21(61),A22(61),A23(61),
     1          G11(61),G12(61),G13(61),G21(61),G22(61),G23(61),
     2          W1(61),W2(61),W3(61),DELF(61),DELU(61)
C - - - - - - - - - - - - - - - - - - - - - - - - - - - - - - - - -
```

```
      W1(1) = R1(1)
      W2(1) = R2(1)
      W3(1) = R3(1)
      A11(1)= 1.0
      A12(1)= 0.0
      A13(1)= 0.0
      A21(1)= 0.0
      A22(1)= 1.0
      A23(1)= 0.0
      G11(2)=-1.0
      G12(2)=-0.5*DETA(1)
      G13(2)= 0.0
      G21(2)= S4(2)
      G23(2)=-2.0*S2(2)/DETA(1)
      G22(2)= G23(2)+S6(2)
      DO 20  J=2,NP
      IF(J .EQ. 2) GO TO 10
      DEN    = (A13(J-1)*A21(J-1)-A23(J-1)*A11(J-1)-A(J)*
     1          (A12(J-1)*A21(J-1)-A22(J-1)*A11(J-1)))
      G11(J)= (A23(J-1)+A(J)*(A(J)*A21(J-1)-A22(J-1)))/DEN
      G12(J)=-(1.0+G11(J)*A11(J-1))/A21(J-1)
      G13(J)= (G11(J)*A13(J-1)+G12(J)*A23(J-1))/A(J)
      G21(J)= (S2(J)*A21(J-1)-S4(J)*A23(J-1)+A(J)*(S4(J)*
     1         A22(J-1)-S6(J)*A21(J-1)))/DEN
      G22(J)= (S4(J)-G21(J)*A11(J-1))/A21(J-1)
      G23(J)= (G21(J)*A12(J-1)+G22(J)*A22(J-1)-S6(J))
   10 A11(J)= 1.0
      A12(J)=-A(J)-G13(J)
      A13(J)= A(J)*G13(J)
      A21(J)= S3(J)
      A22(J)= S5(J)-G23(J)
      A23(J)= S1(J)+A(J)*G23(J)
      W1(J) = R1(J)-G11(J)*W1(J-1)-G12(J)*W2(J-1)-G13(J)*W3(J-1)
      W2(J) = R2(J)-G21(J)*W1(J-1)-G22(J)*W2(J-1)-G23(J)*W3(J-1)
      W3(J) = R3(J)
   20 CONTINUE
      DELU(NP) = W3(NP)
      E1       = W1(NP)-A12(NP)*DELU(NP)
      E2       = W2(NP)-A22(NP)*DELU(NP)
      DELV(NP) = (E2*A11(NP)-E1*A21(NP))/(A23(NP)*A11(NP)-A13(NP)*
     1            A21(NP))
      DELF(NP) = (E1-A13(NP)*DELV(NP))/A11(NP)
      J        = NP
   30 J        = J-1
      E3       = W3(J)-DELU(J+1)+A(J+1)*DELV(J+1)
      DELV(J)  = (A11(J)*(W2(J)+E3*A22(J))-A21(J)*W1(J)-E3*A21(J)*A12(J)
     1         )/(A21(J)*A12(J)*A(J+1)-A21(J)*A13(J)-A(J+1)*
     2            A22(J)*A11(J)+A23(J)*A11(J))
      DELU(J)  =-A(J+1)*DELV(J)-E3
      DELF(J)  = (W1(J)-A12(J)*DELU(J)-A13(J)*DELV(J))/A11(J)
      IF(J .GT. 1) GO TO 30
      WRITE(6, 50 ) V(1,NZ,2),DELV(1)
      DO 40  J=1,NP
      F(J,NZ,2)= F(J,NZ,2)+DELF(J)
      U(J,NZ,2)= U(J,NZ,2)+DELU(J)
   40 V(J,NZ,2)= V(J,NZ,2)+DELV(J)
      U(1,NZ,2)= 0.0
      RETURN
C - - - - - - - - - - - - - - - - - - - - - - - - - - - - - - - - -
   50 FORMAT(1H ,5X,8HV(WALL)=,E14.6,5X,6HDELVW=,E14.6)
      END
```

9.11 Subroutine GROWTH and MAIN

- Subroutine GROWTH allows the boundary-layer thickness to grow as described below
- Subroutine MAIN contains the logic of the computations. Here we also check the convergence of the iterations, using the wall-shear parameter as a test

For most laminar boundary-layer flows, the transformed boundary-layer thickness $\eta_e(X)$ is nearly constant. A value of $\eta_e = 8$ is sufficient. However, for turbulent boundary layers, $\eta_e(X)$ generally increases with increasing X and we may obtain an estimate of $\eta_e(X)$ by the following procedure.

We always require that $\eta_e(X_i) \geqslant \eta_e(X_{j-1})$ and when the computations on $X = X_i$ have been completed, we test to see if $G_j^{i,k} \leqslant 10^{-3}$ in the subroutine MAIN. If this test is satisfied, we set $\eta_e(X_{i+1}) = \eta_e(X_i)$. Otherwise, we call GROWTH and set $J_{\text{new}} = J_{\text{old}} + s$, where s is a number of points, which we take equal to 1. In this case we also specify values of $(f_j^{i,k}, U_j^{i,k}, G_j^{i,k}, b_j^{i,k})$ for the new η points. We take the values of $U_j^{i,k} = U_J^{i,k}, f_j^{i,k} = U_J^{i,k}(\eta_j - \eta_J) + f_J^{i,k}$ and $b_j^{i,k} = b_J^{i,k}$.

The listing for Subroutine GROWTH is given below.

```
      SUBROUTINE GROWTH(LL)
C
      COMMON/BLCO/ NXT,NZT,NX,NZ,NP,NTR,NPT,RL,ETAE,VGP,
     1             A(61),ETA(61),DETA(61)
      COMMON/BLCP/ DELV(61),F(61,61,2),U(61,61,2),V(61,61,2),B(61,61,2)
C  - - - - - - - - - - - - - - - - - - - - - - - - - - - - - - - - -
      NPO   = NP
      NP1   = NP+1
      NPMAX = 61
      IF ( LL .EQ. 1 ) GO TO 95
      NP    = NP + 2
      IF ( NP .GT. NPT ) NP=NPT
      NPMAX = NP
   95 DO 100 J = NP1,NPMAX
      F(J,NZ,2)= U(NPO,NZ,2)*(ETA(J)-ETA(NPO))+F(NPO,NZ,2)
      U(J,NZ,2)= U(NPO,NZ,2)
      V(J,NZ,2)= V(NPO,NZ,2)
      B(J,NZ,2)= B(NPO,NZ,2)
  100 CONTINUE
      RETURN
      END
```

The listing for MAIN is:

```
      COMMON/BLC0/ NXT,NZT,NX,NZ,NP,NTR,NPT,RL,ETAE,VGP,
     1             A(61),ETA(61),DETA(61)
      COMMON/BLC1/ X(41),Z(61),U0(61),P1(61),P2(61),P3(41,61),UE(41,61)
      COMMON/BLCP/ DELV(61),F(61,61,2),U(61,61,2),V(61,61,2),B(61,61,2)
C - - - - - - - - - - - - - - - - - - - - - - - - - - - - - - -
      ITMAX = 10
      NX = 1
      NZ = 1
      CALL INPUT
   10 WRITE (6,110) NX, NZ, X(NX), Z(NZ)
      IF ( NZ .EQ. 1 .AND. NX .GT. 1 ) GO TO 90
      IT    = 0
      IGROW = 0
   20 IT    = IT+1
      IF(IT .LE. ITMAX) GO TO 30
      WRITE(6, 100 )
      STOP
   30 IF(NZ .GE. NTR) CALL EDDY
      IF(NX .GT. 1) GO TO 50
      CALL ICONZ
      GO TO 60
   50 CALL COEFG
   60 CALL SOLV3
C  CHECK FOR CONVERGENCE
      IF(NZ .GE. NTR) GO TO 70
C--LAMINAR FLOW
      IF(ABS(DELV(1)) .GT. 0.0001)GO TO 20
      GO TO 80
C--TURBULENT FLOW
   70 IF(ABS(DELV(1)/(V(1,NZ,2)+0.5*DELV(1))) .GT. 0.02) GO TO 20
   80 IF(NP .EQ. NPT) GO TO 90
      IF(ABS(V(NP,NZ,2)) .LE. 0.001) GO TO 90
      IF(IGROW.GE.2) GOTO 90
      IGROW = IGROW+1
      WRITE(6,120)
  120 FORMAT(1H0,2X,'BOUNDARY LAYER HAS GROWN')
      LL    = 2
      CALL GROWTH(LL)
      GO TO 20
   90 CALL OUTPUT
      GO TO 10
C - - - - - - - - - - - - - - - - - - - - - - - - - - - - - - -
  100 FORMAT(1H0,16X,25HITERATIONS EXCEEDED ITMAX)
  110 FORMAT (1H0,4HNX =,I3,5X,4HNZ =,I3,5X,3HX =,F10.5,5X,3HZ =,F10.5)
      END
```

9.12 Results for unsteady laminar flows with flow reversal

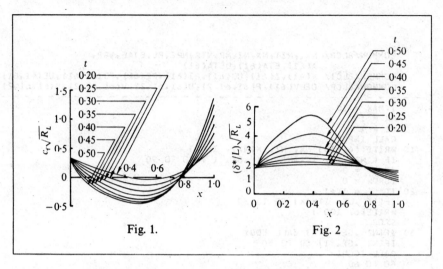

Fig. 1.   Fig. 2

To demonstrate the calculation of unsteady flows with flow reversal by the computer program described in this chapter, we shall use the external velocity distribution given by equations (1) and (2) of panel 9.8. This velocity distribution was used recently by Cebeci (1978) to study the computation of unsteady laminar flows and to see whether there is a singularity associated with such flows. Figures 1 and 2 in the panel show the computed local skin-friction coefficient and the displacement thickness results in $0 < X < 1$ for various values of t with $\alpha = 20$. It is clear from these results that the solution remains smooth for all calculated values of t, even when the region of reversed flow occupies the majority of the boundary layer, and there is no hint of singularity. Instead, as in the study reported by Cebeci (1979), we see the familiar rapid Proudman–Johnson growth (1962) of the boundary layer in the reversed flow region which, unless special measures are taken, terminates the calculations due to the rapid thickening of the boundary layer, and a consequent substantial increase in the number of grid points across the layer.

9.13 Results for unsteady turbulent flows with flow reversal

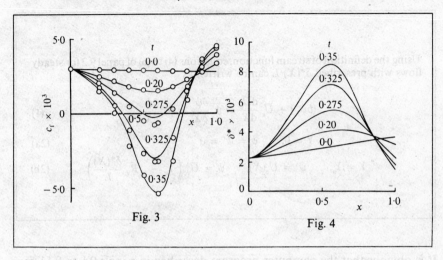

Fig. 3

Fig. 4

Figures 3 and 4 in the panel show the calculations of Cebeci, Carr and Bradshaw, (1979) for turbulent flows with $\alpha = 40$ and a unit Reynolds number $U_\infty/\nu = 2.2 \times 10^6 \, \text{m}^{-1}$. The results in Fig. 3 were obtained by different expressions for the damping-length parameter A; those shown by circles were obtained with the definition of A in panel 3.6, and those shown by solid lines by writing A as

$$A = A^+ \left(\frac{\tau}{\rho}\right)_{\text{max}}^{-1/2}$$

As is seen, both expressions give nearly the same results.

The results in Fig. 4, as in laminar flows, indicate no signs of singularity for any calculated value of t. This is in contrast to the findings of Patel and Nash (1975). Again we see the familiar rapid thickening of the boundary layer in the reversed flow region. If it were not for this, calculations for greater values of t than those considered here could have been made.

9.14 The inverse problem, 1

Using the definition of stream function, equations (4) to (6) of panel 9.2 for steady flows with prescribed $\delta^*(X)/L$ can be written as

$$(b\psi'')' + \bar{U}_e \frac{d\bar{U}_e}{dX} = \psi' \frac{\partial \psi'}{\partial X} - \psi'' \frac{\partial \psi}{\partial X} \tag{1}$$

$$Y = 0, \quad \psi = \psi' = 0 \tag{2a}$$

$$Y = Y_e, \quad \psi' = \bar{U}_e(X), \quad \psi = \bar{U}_e \left(Y_e - \sqrt{R_L} \frac{\delta^*(X)}{L} \right) \tag{2b}$$

It is obvious that the computer program described in panels 9.1 to 9.11 can also be used to solve two-dimensional steady boundary-layer equations for laminar and turbulent flows. This can be done by setting NXT = 1 and by defining P_3 as

$$P_3 = P_2 = \frac{X}{\bar{U}_0} \frac{d\bar{U}_0}{dX},$$

thus reducing the equations to the form for two-dimensional steady flows with the Falkner–Skan transformation.

In this and the following panels, we shall show how this computer program can be used to compute steady, two-dimensional, inverse boundary-layer flows in which the displacement thickness is prescribed and the FLARE approximation is used in the region where U-velocity is negative.

We shall assume that a solution of the equations of panel 9.3 for $T = 0$ is obtained by the present computer program, and that we are interested in obtaining a solution of the system of equations given in panel 7.10. Note that the equations of panel 9.3 are solved for a given external velocity distribution whereas equations (1) and (2) of panel 7.10 are solved for a prescribed displacement thickness distribution. Also, the equations of panel 9.3 use transformed variables while equations (1) and (2) use physical variables.

It is obvious that equations similar to equations (1) and (2) can also be obtained from equations (4) and (5) of panel 9.3. Using the definition of stream function ψ, we can write them and their boundary conditions including the case of specified displacement thickness $\delta^*(X)$, for steady flows in the forms shown in the above panel, which are identical to those in panel 7.10 except that the dP/dX term there is now replaced by the $\bar{U}_e \, d\bar{U}_e/dX$ term.

9.15 The inverse problem, 2: numerical formulation

- Use the Mechul function method. With the function $W(\equiv \bar{U}_e)$ treated as unknown and with the help of the y-momentum equation ($\partial W/\partial Y = 0$), write equations (1) and (2) of panel 9.14 as a system of first-order equations (see panel 7.11 also)
- The finite difference approximations of equations $\psi' = U$ and $U' = G$ are identical to those of equations (2a), (2b) of panel 9.4, except for the notation used for ψ
- The finite difference approximations of the two remaining equations are given below by equations (3a), (3b)

Note that, except for the presence of the W-term, the form of equations (2c) and (3) of panel 9.4 is identical to those of equations (3a), (3b), of this panel if we set $P_1 = P_2 = P_3 = 0$.

$$W_j^i - W_{j-1}^i = 0, \qquad (3a)$$

$$\frac{(bG)_j^i - (bG)_{j-1}^i}{h_{j-1}} + \alpha_i(\psi G)_{j-\frac{1}{2}}^i - \alpha_i(U^2)_{j-\frac{1}{2}}^i + \alpha_i(W^2)_{j-\frac{1}{2}}^i + \alpha_i(G_{j-\frac{1}{2}}^{i-1}\psi_{j-\frac{1}{2}}^i$$

$$- \psi_{j-\frac{1}{2}}^{i-1}G_{j-\frac{1}{2}}^i) = R_{j-\frac{1}{2}}^{i-1}, \qquad (3b)$$

where

$$\alpha_i = \frac{1}{k_i}, \quad R_{j-\frac{1}{2}}^{i-1} = -L_{j-\frac{1}{2}}^{i-1} + \alpha_i[(\psi G)_{j-\frac{1}{2}}^{i-1} - (U^2)_{j-\frac{1}{2}}^{i-1} + (W^2)_{j-\frac{1}{2}}^{i-1}],$$

$$L_{j-\frac{1}{2}}^{i-1} = [(bG)_{j-\frac{1}{2}}']^{i-1}.$$

9.16 The inverse problem, 3: Subroutine ICONZ1

- We next use Newton's method to linearize the nonlinear algebraic equations given by equations (2a), (2b) of panel 9.4 and by equations (3a), (3b) of panel 9.15 together with their boundary conditions. The resulting system can be ordered to be written in the form shown below by equations (4)
- Similarly the boundary conditions can be linearized and written in the form shown below by equations (8) and (10)
- A listing called Subroutine ICONZ1 is presented below. It defines the coefficients of the linear system and will replace ICONZ once we revert to physical variables for $T = 0$

The linear system for $1 \leqslant j \leqslant J$ can be arranged to be written as

$$\delta\psi_j - \delta\psi_{j-1} - \frac{h_{j-1}}{2}(\delta U_j + \delta U_{j-1}) = (r_1)_j, \tag{4a}$$

$$\delta U_j - \delta U_{j-1} - \frac{h_{j-1}}{2}(\delta G_j + \delta G_{j-1}) = (r_3)_{j-1}, \tag{4b}$$

$$\delta W_j - \delta W_{j-1} \qquad\qquad = (r_4)_{j-1} \equiv 0, \tag{4c}$$

$$(S_1)_j \delta G_j + (S_2)_j \delta G_{j-1} + (S_3)_j \delta\psi_j + (S_4)_j \delta\psi_{j-1} + (S_5)_j \delta U_j + (S_6)_j \delta U_{j-1}$$
$$+ (S_7)_j \delta W_j + (S_8)_j \delta W_{j-1} = (r_2)_j. \tag{4d}$$

Here $(r_1)_j$ and $(r_3)_{j-1}$ are given by equations (5a), (5c) of panel 9.4 provided that we set $\psi_j = f_j$. The coefficients $(S_k)_i$ for $k = 1, 2, \ldots, 8$, are also given by the definitions under equation (4c) of the same panel provided that we set P_1 and $P_2 = 0$, and define the parameter $(r_2)_j$ by

$$(r_2)_j = R^{i-1}_{j-\frac{1}{2}} - \{(bG)'_{j-\frac{1}{2}} + \alpha_i(\psi G)_{j-\frac{1}{2}} + \alpha_i[(W^2)_{j-\frac{1}{2}} - (U^2)_{j-\frac{1}{2}}\text{FLARE}]$$
$$+ \alpha_i(G^{i-1}_{j-\frac{1}{2}}\psi_{j-\frac{1}{2}} - \psi^{i-1}_{j-\frac{1}{2}}G_{j-\frac{1}{2}})\}. \tag{5}$$

Here $R^{i-1}_{j-\frac{1}{2}}$ is defined by the expression under equation (3b) of panel 9.15.

Note that in the region where U-velocity is negative, we set the $U\,\partial U/\partial X$ term equal to zero. In the coefficients given above, this is accounted for by multiplying the finite difference approximation of this term by FLARE. If $U_j < 0$, we set FLARE $= 0$; otherwise we set it equal to unity. The terms which contain FLARE, in addition to those above, are

$$(S_5)_j = -\alpha_i U_j \cdot \text{FLARE},$$
$$(S_6)_j = -\alpha_i U_{j-1} \cdot \text{FLARE}, \tag{6}$$
$$R^{i-1}_{j-\frac{1}{2}} = -L^{i-1}_{j-\frac{1}{2}} + \alpha_i\{(\psi G)^{i-1}_{j-\frac{1}{2}} + [(W^2)^{i-1}_{j-\frac{1}{2}} - (U^2)^{i-1}_{j-\frac{1}{2}}\text{FLARE}]\}.$$

The two-wall boundary conditions and the edge boundary conditions follow from equations (2a), (2b). At the wall, since

$$\psi = U = 0 \tag{7}$$

we can write

$$\delta\psi_0 = (r_1)_1 \equiv 0, \tag{8a}$$

$$\delta U_0 = (r_2)_1 \equiv 0. \tag{8b}$$

At the boundary-layer edge, we have

$$U_J = W_J \quad \text{and} \quad \psi_J = W_J \gamma_2, \tag{9}$$

where

$$\gamma_2 = Y_J - \frac{\delta^*}{L}\sqrt{R_L}.$$

In linearized form, equations (9) can be written as

$$\delta\psi_J - \gamma_2 \delta W_J = (r_3)_J = W_J \gamma_2 - \psi_J, \tag{10a}$$

$$\delta U_J - \delta W_J = (r_4)_J = 0. \tag{10b}$$

A listing of this subroutine is given below.

```
      SUBROUTINE ICONZ1
      COMMON/BLC0/ NXT,NZT,NX,NZ,NP,NTR,NPT,INV,RL,ETAE,VGP,GAMMA1,
     1             GAMMA2,A(61),ETA(61),DETA(61),DSD(61)
      COMMON/BLC1/ X(41),Z(61),U0(61),P1(61),P2(61),P3(41,61),UE(41,61)
      COMMON/BLCC/ S1(61),S2(61),S3(61),S4(61),S5(61),S6(61),S7(61),
     1             S8(61),R1(61),R2(61),R3(61),R4(61)
      COMMON/BLCP/ F(61,61,2),U(61,61,2),V(61,61,2),W(61,61,2),
     1             B(61,61,2),DELV(61),DELF(61),DELU(61),DELW(61)
C - - - - - - - - - - - - - - - - - - - - - - - - - - - - - - - - - -
      BEL    = 1.0/(Z(NZ)-Z(NZ-1))
      DO 30  J=2,NP
      FLARE  = 1.0
      IF (U(J,NZ,2) .LT. 0.) FLARE = 0.
C DEFINITION OF AVERAGED QUANTITIES
      FB     = 0.5*(F(J,NZ,2)+F(J-1,NZ,2))
      UB     = 0.5*(U(J,NZ,2)+U(J-1,NZ,2))
      VB     = 0.5*(V(J,NZ,2)+V(J-1,NZ,2))
      FVB    = 0.5*(F(J,NZ,2)*V(J,NZ,2)+F(J-1,NZ,2)*V(J-1,NZ,2))
      USB    = 0.5*(U(J,NZ,2)**2+U(J-1,NZ,2)**2)
      WSB    = 0.5*(W(J,NZ,2)**2+W(J-1,NZ,2)**2)
      DERBV  = (B(J,NZ,2)*V(J,NZ,2)-B(J-1,NZ,2)*V(J-1,NZ,2))/DETA(J-1)
      CFB    = 0.5*(F(J,NZ-1,2)+F(J-1,NZ-1,2))
      CUB    = 0.5*(U(J,NZ-1,2)+U(J-1,NZ-1,2))
      CVB    = 0.5*(V(J,NZ-1,2)+V(J-1,NZ-1,2))
      CFVB   = 0.5*(F(J,NZ-1,2)*V(J,NZ-1,2)+F(J-1,NZ-1,2)*V(J-1,NZ-1,2))
      CUSB   = 0.5*(U(J,NZ-1,2)**2+U(J-1,NZ-1,2)**2)
      CWSB   = 0.5*(W(J,NZ-1,2)**2+W(J-1,NZ-1,2)**2)
      CDERBV = (B(J,NZ-1,2)*V(J,NZ-1,2)-B(J-1,NZ-1,2)*V(J-1,NZ-1,2))/
     1         DETA(J-1)
      CRB    = BEL*(CFVB+(CWSB-CUSB*FLARE))-CDERBV
C
```

```
C     COEFFICIENTS OF THE DIFFERENCED MOMENTUM EQUATION
         S1(J)  = B(J,NZ,2)/DETA(J-1)+0.5*BEL*(F(J,NZ,2)-CFB)
         S2(J)  = -B(J-1,NZ,2)/DETA(J-1)+0.5*BEL*(F(J-1,NZ,2)-CFB)
         S3(J)  = 0.5*BEL*(V(J,NZ,2)+CVB)
         S4(J)  = 0.5*BEL*(V(J-1,NZ,2)+CVB)
         S5(J)  = -BEL*U(J,NZ,2)*FLARE
         S6(J)  = -BEL*U(J-1,NZ,2)*FLARE
         S7(J)  = BEL*W(J,NZ,2)
         S8(J)  = BEL*W(J-1,NZ,2)
C
C     DEFINITIONS OF RJ
         R1(J)  = F(J-1,NZ,2)-F(J,NZ,2)+DETA(J-1)*UB
         R3(J-1)=U(J-1,NZ,2)-U(J,NZ,2)+DETA(J-1)*VB
         R2(J)  =CRB-(DERBV+BEL*FVB-BEL*(WSB-USB*FLARE)-BEL*(CFB*VB-CVB*FB))
         R4(J-1)= 0.
   30 CONTINUE
         R1(1)  = 0.0
         R2(1)  = 0.0
         R3(NP) = W(NP,NZ,2)*GAMMA2-F(NP,NZ,2)
         R4(NP) = 0.
         RETURN
         END
```

9.17 The inverse problem, 4: Subroutine SOLV4

- The linear system given in panel 9.16 can be solved by the block elimination method discussed in panel 5.13. The first two rows of the A_0-matrix contains the two wall boundary conditions and the last two rows of the A_J-matrix contains the two edge boundary conditions

- Subroutine SOLV4 can be used to solve both standard and inverse flows as described below

---

This subroutine contains the recursion formulas that arise in the block elimination method and is similar to Subroutine SOLV3 except that the matrices are 4×4 blocks.

Though it is best to use Subroutine SOLV3 for a standard problem, it is more convenient to use SOLV4 for problems in which standard and inverse methods are required. To show how this can be done, we examine the edge boundary conditions, equations (10a), (10b), and the coefficients of the differenced momentum equation, equations (4d) and (6). We also write the A_J matrix as

$$A_J = \begin{bmatrix} 1 & -h_J/2 & 0 & 0 \\ (S_3)_J & (S_5)_J & (S_1)_J & (S_7)_J \\ \gamma_1 & 0 & 0 & \gamma_2 \\ 0 & 1 & 0 & -1 \end{bmatrix}$$

and set $\gamma_1 = 0$ and $\gamma_2 = 1 \cdot 0$ for standard problems and $\gamma_1 = 1$ and γ_2 as defined for inverse problems.

For the standard problem W is prescribed and we set $(S_7)_j = (S_8)_j = 0$; the rest of the coefficients remain unchanged.

A listing of this subroutine is given below.

```
      SUBROUTINE SOLV4
      COMMON/BLCO/ NXT,NZT,NX,NZ,NP,NTR,NPT,INV,RL,ETAE,VGP,GAMMA1,
     1             GAMMA2,A(61),ETA(61),DETA(61),DSD(61)
      COMMON/BLC1/ X(41),Z(61),UO(61),P1(61),P2(61),P3(41,61),UE(41,61)
      COMMON/BLCC/ S1(61),S2(61),S3(61),S4(61),S5(61),S6(61),S7(61),
     1             S8(61),R1(61),R2(61),R3(61),R4(61)
      COMMON/BLCP/ F(61,61,2),U(61,61,2),V(61,61,2),W(61,61,2),
     1             B(61,61,2),DELV(61),DELF(61),DELU(61),DELW(61)
      DIMENSION A11(61),A12(61),A13(61),A14(61),A21(61),A22(61),A23(61),
     1          A24(61),G11(61),G12(61),G13(61),G14(61),G21(61),G22(61),
     2          G23(61),G24(61),W1(61),W2(61),W3(61),W4(61)
C
C - - - - - - - - - - - - - - - - - - - - - - - - - - - - - - - - -
C
      A11(1)   = 1.0
      A12(1)   = 0.0
      A13(1)   = 0.0
      A14(1)   = 0.0
      A21(1)   = 0.0
      A22(1)   = 1.0
      A23(1)   = 0.0
      A24(1)   = 0.0
      W1(1)    = R1(1)
      W2(1)    = R2(1)
      W3(1)    = R3(1)
      W4(1)    = R4(1)
      DO 10 J = 2,NP
      AA1      = A13(J-1)-A (J)*A12(J-1)
      AA2      = A23(J-1)-A (J)*A22(J-1)
      AA3      = S2(J)-A (J)*S6(J)
      DET      = AA2*A11(J-1)-AA1*A21(J-1)
      AJS      = A (J)**2
      G11(J)   = -(AA2+A21(J-1)*AJS)/DET
      G12(J)   = (A11(J-1)*AJS+AA1)/DET
      G13(J)   = A12(J-1)*G11(J)+A22(J-1)*G12(J)+A (J)
      G14(J)   = A14(J-1)*G11(J)+A24(J-1)*G12(J)
      G21(J)   = (S4(J)*AA2-A21(J-1)*AA3)/DET
      G22(J)   = (A11(J-1)*AA3-S4(J)*AA1)/DET
      G23(J)   = A12(J-1)*G21(J)+A22(J-1)*G22(J)-S6(J)
      G24(J)   = A14(J-1)*G21(J)+A24(J-1)*G22(J)-S8(J)
      A11(J)   = 1.0
      A12(J)   = -A (J)-G13(J)
      A13(J)   = A (J)*G13(J)
      A14(J)   = -G14(J)
      A21(J)   = S3(J)
      A22(J)   = S5(J)-G23(J)
      A23(J)   = S1(J)+A (J)*G23(J)
      A24(J)   = S7(J)-G24(J)
      W1(J)    = R1(J) -G11(J)*W1(J-1)-G12(J)*W2(J-1)-W3(J-1)*G13(J)
     1           -G14(J)*W4(J-1)
      W2(J)    = R2(J) -G21(J)*W1(J-1)-G22(J)*W2(J-1)-W3(J-1)*G23(J)
     1           -G24(J)*W4(J-1)
      W3(J)    = R3(J)
      W4(J)    = R4(J)
   10 CONTINUE
      D = GAMMA1*(A13(NP)*A24(NP)-A14(NP)*A23(NP)-A12(NP)*A23(NP)+
     1     A13(NP)*A22(NP)) + GAMMA2*(A11(NP)*A23(NP)-A13(NP)*A21(NP))
```

```
      DF = W3(NP)*(A13(NP)*A24(NP)-A14(NP)*A23(NP)-A12(NP)*A23(NP)+
     1        A13(NP)*A22(NP)) - GAMMA2*(W4(NP)*(A12(NP)*A23(NP)-A13(NP)*
     2        A22(NP)) - W1(NP)*A23(NP)+W2(NP)*A13(NP))
      DU = W4(NP)*(GAMMA1*(A24(NP)*A13(NP)-A23(NP)*A14(NP)) + GAMMA2*
     1        (A11(NP)*A23(NP)-A21(NP)*A13(NP)))+ GAMMA1*(W2(NP)*A13(NP)-
     2        W1(NP)*A23(NP)) - W3(NP)*(W1(NP)*A21(NP)-A11(NP)*W2(NP))
      DV = GAMMA1*(W1(NP)*A24(NP)-W2(NP)*A14(NP)) + W3(NP)*(A21(NP)*
     1        A14(NP)-A11(NP)*A24(NP)) - GAMMA2*(W1(NP)*A21(NP)-W2(NP)*
     2        A11(NP)) +W4(NP)*(GAMMA1*(A22(NP)*A14(NP)-A24(NP)*A12(NP))
     3        +GAMMA2*(A21(NP)*A12(NP)-A11(NP)*A22(NP)))+ GAMMA1*(W1(NP)*
     4        A22(NP)-W2(NP)*A12(NP)) + W3(NP)*(A21(NP)*A12(NP)-A11(NP)*
     5        A22(NP))
      DW = GAMMA1*(W2(NP)*A13(NP)-W1(NP)*A23(NP)+W4(NP)*(A12(NP)*A23(NP)
     1        -A13(NP)*A22(NP))) + W3(NP)*(A11(NP)*A23(NP)-A13(NP)*
     2        A21(NP))
C
      DET      = D
      DELF(NP) = DF/DET
      DELU(NP) = DU/DET
      DELV(NP) = DV/DET
      DELW(NP) = DW/DET
      J        = NP
   20 J        = J-1
      CC1      = DELU(J+1)-W3(J)-A (J+1)*DELV(J+1)
      CC2      = DELW(J+1)-W4(J)
      CC3      = A13(J)-A (J+1)*A12(J)
      CC4      = W1(J)-A12(J)*CC1-A14(J)*CC2
      CC5      = A23(J)-A (J+1)*A22(J)
      CC6      = W2(J)-A22(J)*CC1-A24(J)*CC2
      DENO     = A11(J)*CC5-A21(J)*CC3
      DELF(J)  = (CC4*CC5-CC3*CC6)/DENO
      DELV(J)  = (A11(J)*CC6-A21(J)*CC4)/DENO
      DELW(J)  = CC2
      DELU(J)  = CC1-A (J+1)*DELV(J)
      IF(J .GE. 2) GO TO 20
      WRITE (6,9000) V(1,NZ,2),DELV(1)
      DO 30 J = 1,NP
      F(J,NZ,2) = F(J,NZ,2)+DELF(J)
      U(J,NZ,2) = U(J,NZ,2)+DELU(J)
      V(J,NZ,2) = V(J,NZ,2)+DELV(J)
      W(J,NZ,2) = W(J,NZ,2)+DELW(J)
   30 CONTINUE
      U(1,NZ,2) = 0.
      RETURN
C - - - - - - - - - - - - - - - - - - - - - - - - - - - - - - - -
 9000 FORMAT(1H ,3X,6HV(1) =,E14.6,2X,6HDELV1=,E14.6)
      END
```

10 Recirculating Flows

10.1 Purposes and outline of chapter

- To outline the main features of numerical procedures for the solution of the equations appropriate to recirculating flows with large regions of separation
- To indicate the differences between available procedures and some of the relative advantages
- To present a sample of available results which allow an assessment of the value of solution methods for elliptic equations

Two consequences of closed streamlines in viscous flows are that diffusion is important in two directions and that all regions of the flow can influence all other regions. Thus, for example, the flow immediately behind a step is influenced by the geometry in the downstream region and the region of separated flow near the trailing edge of a wing at angle of attack is influenced by the shape of the trailing edge. If the region of separation is small, boundary-layer assumptions may approximate the flow and the inverse procedure of Chapter 7 can be used. This is, however, an approximation and the flow is correctly represented by equations which include pressure gradients and diffusion in more than one direction.

This chapter is concerned with methods for solving the elliptic equations which characterize recirculating flows. The introduction to the corresponding numerical methods will be concerned with two-dimensional equations and flows but the arguments can readily be extended to three dimensions as is demonstrated in the examples presented in the closing panels.

10.2 Equations and implications, 1

- $\dfrac{\partial \Phi}{\partial t} + U \dfrac{\partial \Phi}{\partial x} + V \dfrac{\partial \Phi}{\partial y} + S_\Phi - \left(\dfrac{\partial}{\partial x}\left[\Gamma_\Phi \dfrac{\partial \Phi}{\partial x} \right] + \dfrac{\partial}{\partial y}\left[\Gamma_\Phi \dfrac{\partial \Phi}{\partial y} \right] \right) = 0$

- where $\Phi = U, V, W, T, m_j, \overline{u_i u_j}$, etc.

    S = source term, for example $\dfrac{1}{\rho} \dfrac{\partial P}{\partial x}$

    and Γ_Φ = diffusion coefficient

- Boundary conditions for each dependent variable or its gradient required on all sides of the solution domain

The equation shown in the panel characterizes two-dimensional, unsteady recirculating flows with constant properties. In contrast to the corresponding boundary-layer equations considered in previous lectures, it contains two diffusion terms. The steady-state form of this equation is elliptic and boundary conditions are, as a result, required on all sides of the geometric solution domain. Again, this may be contrasted with boundary-layer equations whose parabolic nature requires a solution domain described by boundary conditions on three sides. An important implication for numerical procedures is that the marching approach which characterizes the solution of boundary-layer equations is not valid for the recirculating-flow equations where iteration is required.

In the particular case of the momentum equations Φ will represent U and V, for example, and the corresponding values of S_Φ will be $(1/\rho)(\partial P/\partial x)$ and $(1/\rho)(\partial P/\partial y)$. This implies three independent variables and, with mass continuity, three equations so that solutions are possible. For turbulent flow, the diffusion terms will include terms of the form $(\partial/\partial x_j)\,\overline{u_i u_j}$ and these may influence the form of the resulting equation depending on the corresponding turbulence model assumptions. In practice, these terms often have the same form as the laminar diffusion terms; in other cases the source term may be influenced.

10.3 Equations and implications, 2

- $a_\Phi \left[\dfrac{\partial \Phi}{\partial x} \dfrac{\partial \psi}{\partial y} - \dfrac{\partial \Phi}{\partial y} \dfrac{\partial \psi}{\partial x} \right] + S_\Phi - \dfrac{\partial}{\partial x}\left[b_\Phi \dfrac{\partial}{\partial x}(c_\Phi \Phi) \right] - \dfrac{\partial}{\partial y}\left[b_\Phi \dfrac{\partial}{\partial y}(c_\Phi \Phi) \right] = 0$
- where $\Phi = \omega, \psi, T, m_j,$ etc.
- for $\quad \Phi = \omega$ (vorticity), $a_\Phi = b_\Phi = 1, c_\Phi = \mu, S_\Phi = 0$
- for $\quad \Phi = \psi$ (stream function), $a_\Phi = 0, b_\Phi = 1/\rho, c_\Phi = 1, S = -\omega$

The previous panel implied the use of "primitive" variables. An alternative approach is to rewrite the momentum equations in terms of vorticity and the stream function and this is indicated on the panel. An immediate advantage is that only two differential equations need be solved with two dependent variables. Thus, the continuity equation is satisfied by the definition of the stream function and the pressure terms have been eliminated by cross differentiation. This approach was recommended by Gosman et al. (1969) but was subsequently dropped in favour of primitive variables.

The arrangement has disadvantages which appear to be greater than the advantages of the elimination of pressure and the implicit incorporation of continuity. The boundary conditions are less easy to specify and the dependent variable not always so readily understood. It is usually the velocity values which are required and integration is necessary to obtain them. In addition, and perhaps of crucial importance, the extension of the stream-function, vorticity approach to nonuniform density and/or three dimensions poses formidable problems.

Recirculating Flows 195

10.4 Equations and implications, 3

- The selection of a coordinate system can greatly influence the precision of solution and has consequences for the differential equations
- Eulerian and Lagrangian coordinate systems are possible
- Eulerian systems are more common and may be orthogonal, nonorthogonal, specific, general curvilinear, transformed etc.

Rectangular Cartesian coordinates are the most frequently found in the engineering literature and the equations of the previous two pages are written in this form. It can readily be appreciated that cylindrical and spherical coordinates may be more appropriate for some flows and simple geometric transformations allow the corresponding equations to be obtained. The stream function can be a useful independent variable in some cases. It is also possible to rewrite the equations in coordinates which move arbitrarily in space or move with the fluid (Lagrangian). The general curvilinear form of the equations has been considered by Bryant and Humphrey (1976); reduced forms of the equations are also discussed in the paper. They have been used, for example, by Humphrey (1978).

The generation of grid arrangement, both orthogonal and nonorthogonal, is a necessary task in many flow configurations, if precise solutions to the differential equations are to be obtained. One possible method is to solve Laplace's equation for Cartesian coordinates and to use the resulting stream lines as the basis for a coordinate system for the viscous equations. Related work has been reported by Pope (1977a) and others.

10.5 Algebraic equations, 1

- Algebraic equations, obtained from the partial differential equations, may satisfy conservation within each discrete subdivision and must satisfy conservation over the solution domain
- If $S_\Phi = 0$ ($\Phi \neq U, V$), the value of Φ must everywhere be bounded by the maximum and minimum values
- These algebraic equations may be formulated from finite difference, finite element and finite volume assumptions which can be identical

The elliptic differential equations, like the parabolic equations previously discussed, cannot be solved analytically except in a few special cases and numerical methods are required. Thus, algebraic equations are formed to represent the flow in a large number of discrete cells and are solved simultaneously. The algebraic equations must satisfy the two conditions of the panel.

The algebraic equations may be formulated in different ways and two of these are indicated in the following two panels. The arguments are closely related to those of Chapter 5 but here the possible influence of all regions of a flow on the properties at a specific location and the iterative nature of the solution must be borne in mind although the main differences occur in the solution algorithm. An additional difference relates to pressure which, unlike boundary layer flows, is determined from the calculation.

The relative merits of finite difference, finite element and finite volume assumptions are, in some respect, a matter of argument in relation to fluid-mechanics. It is clear that they can be equivalent and also that, in some form, significant practical differences can occur. At present, finite element methods have been developed largely for solid-mechanics problems where nonrectangular mesh geometries can be useful; their direct translation to fluid mechanics problems, with their greater nonlinear contributions, usually leads to Reynolds number limitations.

10.6 Algebraic equations, 2

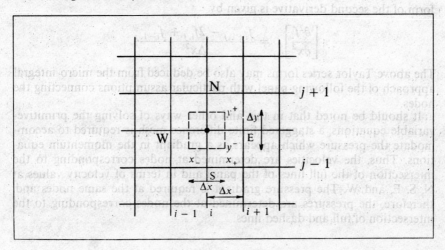

The formulation of finite difference equations is similar to that described previously in connection with boundary-layer equations (Chapter 5), although the accuracy of the resulting algebraic equations will depend on the relationship of the direction of streamlines to the grid. Thus, the differencing discussed briefly below and expressed in terms of five nodes may result in errors which make desirable the use of more complicated schemes, such as the nine-point differencing discussed in relation to panel 10.8.

The use of truncated Taylor series expansions leads to

$$f_{1+i,j} = f_{i,j} + \left[\frac{\partial f}{\partial x}\right]_{i,j}(x_{i+1,j} - x_{i,j}) + \frac{1}{2}\left[\frac{\partial^2 f}{\partial x^2}\right]_{i,j}(x_{i+1,j} - x_{i,j})^2 + \ldots$$

$$= f_{i,j} + \left[\frac{\partial f}{\partial x}\right]_{i,j}\Delta x + \frac{1}{2}\left[\frac{\partial^2 f}{\partial x^2}\right]_{i,j}\Delta x^2 + \ldots$$

and

$$\left[\frac{\partial f}{\partial x}\right]_{i,j} = \frac{f_{i+1,j} - f_{i,j}}{\Delta x} + \frac{1}{2}\left[\frac{\partial^2 f}{\partial x^2}\right]_{i,j}\Delta x + \ldots$$

$$= \frac{f_{i+1,j} - f_{i,j}}{\Delta x} + \text{terms of order } \Delta x \text{ and smaller.}$$

This forward difference approximation is first-order accurate with an error of order Δx. The central difference approximation,

$$\left[\frac{\partial f}{\partial x}\right]_{i,j} = \frac{f_{i+1,j} - f_{i-1,j}}{2\Delta x}$$

is second-order accurate with an error of order Δx^2. The central difference form of the second derivative is given by

$$\left[\frac{\partial^2 f}{\partial x^2}\right]_{i,j} = \frac{f_{i+1,j} - 2f_{i,j} + f_{i-1,j}}{\Delta x^2}.$$

The above Taylor series forms may also be deduced from the micro-integral approach of the following panel with particular assumptions connecting the nodes.

It should be noted that in this and other ways of solving the primitive-variable equations, a staggered finite difference mesh is required to accommodate the pressure which appears as a gradient in the momentum equations. Thus, the velocities are determined at nodes corresponding to the intersection of the full lines of the panel and in terms of velocity values at N, S, E, and W. The pressure gradient is required at the same nodes and, therefore, the pressures are determined at the nodes corresponding to the intersection of full and dashed lines.

10.7 Algebraic equations, 3

- Integrate differential equations, $\int_{x_-}^{x_+} \int_{y_-}^{y_+} dy\, dx$

  i.e. $\dfrac{\partial}{\partial t} \iint \rho\Phi\, dy\, dx + \left\{ \int_{y_-}^{y_+} \left(\rho U\Phi - \Gamma_\Phi \dfrac{\partial \Phi}{\partial x} \right) dy \right\}_{x_-}^{x_+}$
  $+ \left\{ \int_{x_-}^{x_+} \left(\rho V\Phi - \Gamma_\Phi \dfrac{\partial \Phi}{\partial y} \right) dx \right\}_{y_-}^{y_+} = \iint S_\Phi\, dy\, dx$

- Provided finite difference approximations are applied consistently, the resulting scheme is conservative

---

As described by Roache (1972a) and others, finite difference equations may be obtained by fitting analytical expressions, for example, polynomials, to link the node points; the coefficients of the polynomials then appear in the finite difference equations. The first-order differentials evaluated in this way will be identical to the central difference expressions of the previous panel if straight lines are chosen and the nodes are uniformly distributed.

The micro-integral method has been used, for example by Patankar and Spalding (1970, 1972) and Gosman and Pun (1973) and leads, for the control volume of the previous panel, to the integral equation shown above; corresponding finite difference equations are discussed on the following page. A closely related procedure is the PIC (particle in cell) approach used by the Los Alamos group, see for example Gentry *et al.* (1966). It give similar results to the micro-integral approach.

10.8 Algebraic equations, 4

The problem: 5-point representations of

$$\frac{\partial}{\partial x}(\rho U\Phi) + \frac{\partial}{\partial y}(\rho V\Phi) - \Gamma\left(\frac{\partial^2 \Phi}{\partial x^2} + \frac{\partial^2 \Phi}{\partial y^2}\right) = 0$$

have the general form:

$$\Phi_2 = a_1\Phi_1 + a_3\Phi_3 + a_4\Phi_4 + a_5\Phi_5$$

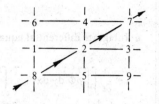

But, when the flow is inclined at, e.g. 45° to the mesh, the correct solution is

$$\Phi_2 = \Phi_8$$

Result is "smearing" of Φ profile across grid

The combination of small numbers of nodes and streamlines diagonal to the cell can lead to errors and particularly to "false" or numerical diffusion. This is a fundamental problem of numerical methods and can be reduced by introducing more nodes, increasing the "order" of representation by involving points 6, 7, 8 and 9 in the finite difference representation or by aligning the cell with the stream line.

If the number of nodes is too small so that the cell Reynolds (Peclet) number becomes greater than 2, the diagonal dominance of the matrix will be lost with consequent problems for the solution algorithm. In such cases, assumptions such as those of upwind or hybrid differencing may be required to allow a solution to be obtained. This topic is discussed further in relation to panel 10.11 but it should be realized that, in some flows and particularly where three-dimensional equations are solved, limitations of computer storage capacity may imply that some cell Reynolds numbers are very much larger than 2. When this occurs, upwind differencing is required and, if the streamline is diagonal as shown in the panel, there will be consequent numerical diffusion and possible large error. Nine-point differencing can, in principle, reduce the error but may give convergence problems which preclude a solution. There is no doubt that further research is necessary to develop schemes which avoid or reduce numerical diffusion and, where it is inevitable, to quantify its effect.

10.9 Solution algorithms

- The finite difference equations reduce to a set of algebraic equations for the dependent variables U_i, P, k, ε etc.
- The solution algorithm may use simultaneous, successive or hybrid procedures
- Simultaneous procedures involve iterative solutions of the equations over the entire mesh or specified portions: see, for example, Weinstein et al. (1969)
- Successive procedures separate the algebraic equations into sub-matrices for each variable and the successive solutions are repeated since the coefficients depend on other variables: see, for example, Gosman et al. (1969)
- Hybrid procedures make use of simultaneous and successive solutions in separate regions of the solution domain: see, for example, Briley and McDonald (1973)

The relative advantages of the possibilities of the panel will depend on the particular problem. The successive approach appears to offer the advantage of minimum computer storage requirements but is susceptible to numerical instabilities. The simultaneous approach allows all linear interactions to be represented simultaneously but can have much larger storage requirements. In general, therefore, hybrid approaches and the flexibility which they imply are required and make use of block solution methods rather than the line-by-line and point iteration schemes found in the other approaches.

Two such approaches are the ADI (alternating direction implicit) and pressure-corrector method. The former was introduced by Peaceman and Rachford (1965) and Douglas (1955) and may be contrasted with the ADE (alternating direction explicit) methods of Saul'yev (1964) and Richtmyer and Morton (1967). It may be summarized by

$$\frac{\Phi^{n+\frac{1}{2}} - \Phi^n}{\frac{1}{2}\Delta t} = \frac{\partial}{\partial x}\left\{\Gamma_\Phi \frac{\partial \Phi}{\partial x} - \rho U \Phi\right\}^{n+\frac{1}{2}} + \frac{\partial}{\partial y}\left\{\Gamma_\Phi \frac{\partial \Phi}{\partial y} - \rho V \Phi\right\}^n,$$

$$\frac{\Phi^{n+1} - \Phi^{n+\frac{1}{2}}}{\frac{1}{2}\Delta t} = \frac{\partial}{\partial x}\left\{\Gamma_\Phi \frac{\partial \Phi}{\partial x} - \rho U \Phi\right\}^{n+\frac{1}{2}} + \frac{\partial}{\partial y}\left\{\Gamma_\Phi \frac{\partial \Phi}{\partial y} - \rho V \Phi\right\}^{n+1}.$$

- Solve first equation along $y = $ const. lines using tridiagonal matrix algorithm (TDMA) as in panel 5.9.
- Solve second equation along $x = $ const. lines using tri-diagonal matrix algorithm.
- Repeat for next time step and continue until $\Phi^{n+1} = \Phi^n$.

The procedure reported by Patankar and Spalding (1972), for example, makes use of an algorithm which is based on the successive solution of equations for the dependent variables. It begins with an estimated pressure field and calculates values of the velocity components etc. which, in general, do not satisfy continuity. Pressure corrections are then calculated at each finite-difference node to satisfy continuity and are added to the estimated pressure field to allow a new solution for the dependent variables. This process is repeated until an acceptable convergence, denoted by a maximum residual error for each variable, is achieved. The finite difference equations are solved by a line-by-line application of the TDMA. In three-dimensional problems line-by-line solution can be applied in one plane and plane-by-plane solution in the remaining direction.

Recirculating Flows

10.10 Classification of some methods

- Gosman *et al.* (1969). Two-dimensional; stream function-vorticity; compressible; finite volume, implicit; successive, point iteration

- Briley and McDonald (1974). Two-dimensional; stream function-vorticity; compressible; finite difference, implicit, artificial viscosity; successive, line (ADI) iteration

- Harlow and Amsden (1971). Three-dimensional; primitive variable; compressible; finite volume, partially implicit, upwind; successive point iteration

- Patankar and Spalding (1972). Three-dimensional; primitive variable; compressible; finite volume, implicit, upwind; successive line iteration

There are many methods available for the solution of elliptic equations and those of the panel have been selected to allow comment on some of the differences. The first two make use of the stream function and vorticity discussed in related to panel 10.3 and the second two the primitive variables of panel 10.2. The latter are undoubtedly more general and allow convenient extensional flows. The former approach is not conveniently extended to three-dimensional flows although the vorticity transport equation can (see, for example, Bower and Peters, 1979), but can have significant advantages for some flows. Largely to allow the calculation of three-dimensional flows the method of Gosman *et al.* (1969) was abandoned at Imperial College in favour of the primitive-variable approach incorporated in the SIMPLE algorithm of Patankar and Spalding (1972) and the TEACH program of Gosman and Pun (1973). Similarly, Briley, McDonald and Gibeling (1977) developed a method with velocity and pressure as dependent variables and, as with several methods and their application to complex problems, have tended to make use of hybrid successive/simultaneous, block iteration algorithms.

It is useful to note the almost exclusive use of implicit numerical schemes which stems from their second-order accuracy and stability. This requires modification if solutions are required at large cell Reynolds numbers and upwind differencing has been used in the last two methods of the panel. The method of Briley and McDonald has the same problem and overcomes it in a related manner which again amounts to the inclusion of artificial viscosity.

A useful comparison of several methods, in the particular context of combusting flows, has been presented by McDonald (1978).

10.11 Additional considerations

- Implementation of boundary conditions
- Irregular and nonrectangular boundaries
- Reynolds stress models
- Compressible flows
- Grid refinement, convergence and numerical diffusion

The literature relating to the solution of the equations referred to here is extensive and the present material is not intended to provide more than an indication of alternative methods. The five topics of the panel can be important and are discussed briefly in the following paragraphs.

Numerical methods, with their discrete node points, require that values of Φ at the node adjacent to the wall be related to wall values, for example, shear stress. The simplest and most correct procedure is to relate the wall shear stress to the laminar-flow viscosity and the velocity gradient in the viscous sublayer. The need for economy of grid nodes renders this impractical and law-of-the-wall procedures are necessary. It should be stressed that, although local regions of flow can and in some cases should be calculated with comparatively fine grids, most practical flow problems require that the complete flow field be represented by a small number of carefully located nodes. A useful example is the gas turbine combustor and the need to investigate the flow characteristics in the neighbourhood of the fueling device and primary zone: this cannot be achieved without knowledge of the downstream boundary condition. One approach to this problem is to calculate, with a small number of grid nodes, the flow throughout the combustor with $\partial \Phi / \partial x = 0$ in the exit plane; this calculation provides answers at the exit from the primary zone which may be used as downstream boundary conditions for primary-zone calculations with grid nodes concentrated in this region.

Irregular and nonrectangular boundaries do not, in principle, present difficulties for numerical solution methods. Small changes in dimensions with, perhaps, disproportionately large changes in flow patterns, require careful location of a large number of nodes with consequent expense. It is also easy to appreciate that rectangular grid arrangements can be uneconomic in some regions of flow and imprecise in others. In the latter case the use of nonrectangular grids may be desirable and can be accommodated by the use of nonorthogonal grids, though at the expense of solving complicated and unwieldy equations. It is often preferable to generate an orthogonal grid by solving Laplace's equation as indicated previously.

For some flows, for example wall-jet arrangements and annular-type flows, a Reynolds stress model of turbulence may be required. As indicated by Pope and Whitelaw (1976), the $\overline{u_i u_j}$ values have to be located at nodes other than those used for U_i and P with added numerical complexity. They also showed that, at least for two bluff-body flows, the use of stress models does not lead to significantly better results as indicated in relation to panel 10.13. This result may stem in part from inaccuracies in the computational method.

Compressible flows require special consideration even at subsonic Mach numbers where shocks are unimportant. The pressure-correction equations, for example, require modification to account for variations in density and their influence on pressure and velocity.

Lastly, it should be clear that optimization of numerical procedures is a difficult if not impossible task. Calculations have to be performed with different numbers of grid nodes to attempt to ensure that a converged and correct solution has been achieved. Numerical diffusion is, in general, reduced by increasing the number of grid nodes though at increased cost. It has recently been shown at Imperial College that calculations with 39×39 nodes lead to values of mean velocity more than 5 per cent different from those obtained with 20×20 nodes. This result was obtained for the flow in the comparatively simple geometry of an air-conditioned room by Restivo (1979). Similar numbers have been obtained in furnace geometries. The obvious question of what result is obtained with, say, 100×100 nodes remains to be answered. The papers by Castro (1979) and Militzer *et al.* (1977) raise similar questions and only partial answers are available. Relevant comments, based on experience, are provided in the following pages. Leonard (1979) provides an amusing and thought-provoking commentary again with partial answers.

10.12 Examples of calculations of flows with recirculation

- The following panels present results obtained with a typical solution procedure for elliptic equations. In each case, comments are included to indicate the computer requirements
- The results were obtained with the TEACH program developed at Imperial College
- The examples have been chosen to indicate what can be done, with what precision and at what cost. They include calculations of relevance to the topics of Chapters 11, 14 and 15

The previous panels have briefly reviewed methods of solving elliptic equations with emphasis on two-dimensional systems. The remainder of the lecture is devoted to demonstrating what can be achieved with one of these methods and, in particular, that developed and used at Imperial College by Gosman, Lockwood, Whitelaw and their co-workers. It makes use of a pressure-corrector algorithm and is based on the work of Gosman and Pun (1973). The range of examples is small, due to space limitations: further recent examples have been described, for example, those of Gosman and Watkins (1978), Gosman et al. (1978), Gosman et al. (1979), Humphrey (1978), Nielsen et al. (1978), and Habib and Whitelaw (1979, 1980). It has been chosen to allow reference to most of the topics of the previous panel and to include topics of importance in future chapters. Thus, it includes:

a. Disc stabilized flows: two-dimensional (axisymmetric).
b. Coaxial confined jet flows: two-dimensional (axisymmetric).
c. Square-duct bend: three-dimensional.
d. Round jet in cross flow: three-dimensional.

The number of independent variables appropriate to each example is indicated. In the case of c, results have been obtained with a fully elliptic, three-dimensional procedure and are compared with published results obtained with reduced forms of these equations, in which longitudinal diffusion terms are omitted in one case and a cross-section averaged pressure used in the equation for the longitudinal component of momentum, together with the neglect of longitudinal diffusion, in the other.

10.13 Disc-stabilized flow

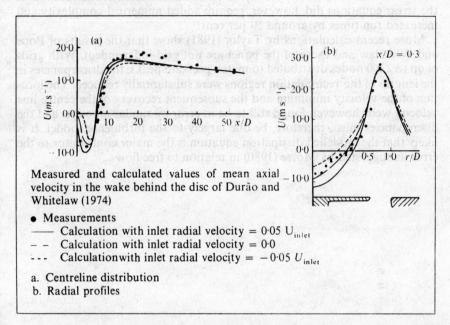

Measured and calculated values of mean axial velocity in the wake behind the disc of Durão and Whitelaw (1974)

- Measurements
- ——— Calculation with inlet radial velocity = 0·05 U_{inlet}
- – – Calculation with inlet radial velocity = 0·0
- - - - Calculation with inlet radial velocity = $-0·05\ U_{inlet}$

a. Centreline distribution
b. Radial profiles

The two figures of the panel were obtained with a two-equation model of turbulence, with approximately 20 × 20 node points, around 30 000 words of CDC6600 store and run times of around 200 s. The boundary conditions corresponded to the measurements of Durão and Whitelaw (1974) in the plane of the disc, to symmetry along the geometric axis, to zero velocity on the walls of the confining pipe and on the disc surface, and to zero gradients far downstream. As can be seen, the general form of the calculated results is correct but the recirculation length is too short. The influence of the radial velocity profile in the plane of the disc is demonstrated. The later measurements of Durão and Whitelaw (1977) show that the initial radial-velocity profile is not uniform and has local values greater than 0·07 U. It is unlikely that the inclusion of the measured V-profiles would result in correct calculations in view of the variations indicated in the panel.

Pope and Whitelaw (1976) reported similar calculations and made comparisons with wakes, without recirculation, and with the disk-stabilized, recirculating flow of Carmody (1964). The nonrecirculating wakes were accurately predicted: the data of Carmody were represented with precision similar to that of the panel and the discrepancies are again due in part to the numerical procedure and in part to the turbulence model. They also compared their two-equation, turbulence model results with those obtained with stress models corresponding to those of Launder *et al.* (1975) and Naot *et al.* (1973). The use of the stress models did not result in significant changes in calculated values of mean velocity or turbulence intensity. The solution of the stress equations did, however, require added numerical complexity and increased run times by around 50 per cent.

More recent calculations by Taylor (1981) show that the results of Pope and Whitelaw and those of the panel are not grid-independent. With grids of up to 2200 nodes distributed to match pressure peaks, the discrepancies in the lengths of the recirculation regions were substantially reduced. The location of the velocity minimum and the subsequent recovery of the centre-line velocity were, however, found still to be in error by similar amounts and the discrepancies must, therefore, be due largely to the turbulence model. It is likely that the modelled dissipation equation is the major contributor to the error as suggested by Morse (1980) in relation to free flows.

10.14 Coaxial, confined jet flow

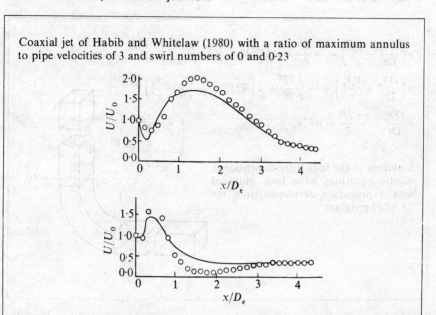

Coaxial jet of Habib and Whitelaw (1980) with a ratio of maximum annulus to pipe velocities of 3 and swirl numbers of 0 and 0·23

The results of the panel correspond to calculations obtained with a two-equation model of turbulence and computer storage and run times slightly longer than those associated with the previous panel, especially for the case of the swirl number of 0·23. The measurements were obtained by a combination of hot-wire and laser-Doppler anemometry. Previous calculations, obtained with a stress model for the zero swirl case, were reported by Habib and Whitelaw (1979) and were only slightly different.

The result for zero swirl reveal discrepancies which can probably be associated with the two-equation turbulence model. The maximum cell Reynolds number associated with the calculations was less than 2 and, although numerical errors undoubtedly occur, particularly in the upstream region, they are probably less than those associated with the model. It can be expected that the larger discrepancies in the finite swirl case will be reduced by the use of a stress model. The algebraic stress model results of Gibson (1978) for curved flows suggest that this approach would also lead to improvement although a full stress model may be preferable. Bradshaw (1973) has shown that eddy viscosity models are not satisfactory for swirling flows and Morse (1980) found it necessary to make the dissipation equation constant $c_{\varepsilon 1}$ a function of Richardson number to represent his free swirling-jet flows.

10.15 Square-duct bend, 1

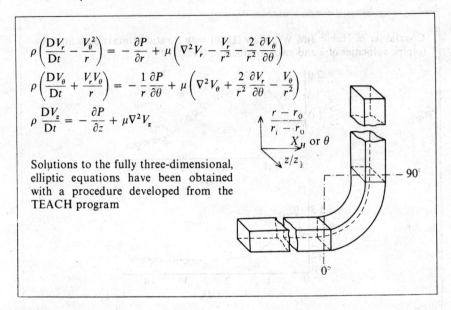

$$\rho\left(\frac{DV_r}{Dt} - \frac{V_\theta^2}{r}\right) = -\frac{\partial P}{\partial r} + \mu\left(\nabla^2 V_r - \frac{V_r}{r^2} - \frac{2}{r^2}\frac{\partial V_\theta}{\partial \theta}\right)$$

$$\rho\left(\frac{DV_\theta}{Dt} + \frac{V_r V_\theta}{r}\right) = -\frac{1}{r}\frac{\partial P}{\partial \theta} + \mu\left(\nabla^2 V_\theta + \frac{2}{r^2}\frac{\partial V_r}{\partial \theta} - \frac{V_\theta}{r^2}\right)$$

$$\rho\frac{DV_z}{Dt} = -\frac{\partial P}{\partial z} + \mu\nabla^2 V_z$$

Solutions to the fully three-dimensional, elliptic equations have been obtained with a procedure developed from the TEACH program

The equations shown on the panel are appropriate to laminar flow in a curved duct of rectangular cross-section. Humphrey et al. (1977) solved equations of this form in order to calculate the low Reynolds number flow of water in a 90° bend of 40 mm × 40 mm cross-section and of a mean radius of 92 mm. In their case, the bend was located downstream of a 1·8 m and upstream of a 1·2 m straight section of the same duct.

The equations are written in cylindrical coordinates and correspond to a steady-state flow. Since it is possible, and indeed has been demonstrated, that the flow in the bend can recirculate, further simplifications have not been incorporated. Thus,

$$\frac{D}{Dt} = V_r\frac{\partial}{\partial r} + \frac{V_\theta}{r}\frac{\partial}{\partial \theta} + V_z\frac{\partial}{\partial z}$$

and

$$\nabla^2 = \frac{\partial^2}{\partial r^2} + \frac{1}{r}\frac{\partial}{\partial r} + \frac{1}{r^2}\frac{\partial^2}{\partial \theta^2} + \frac{\partial^2}{\partial z^2}.$$

Of course, when the radius of curvature is infinite, the momentum and continuity equations reduce to the more familiar rectangular form

$$\frac{\partial U_i}{\partial x_i} = 0,$$

$$U_j\frac{\partial U_i}{\partial x_j} = -\frac{\partial P}{\partial x_i} + \frac{\partial}{\partial x_j}\left(\mu\frac{\partial U_i}{\partial x_j}\right).$$

Recirculating Flows 211

The boundary conditions used by Humphrey *et al.* (1977) were:
initial plane (all z and r at $\theta = X_H = -10$):

$V_\theta =$ developed duct flow
$V_z = V_r = 0$

sidewalls (all θ or X_H):

$V_\theta = V_r = V_z = 0$ at $z = \pm z_{\frac{1}{2}}$
and $r = r_o$ and r_i

symmetry plane (all r and θ or X_H at $z = 0$)

$$V_z = 0$$

$$\frac{\partial V_\theta}{\partial x} = \frac{\partial V_r}{\partial z} = 0$$

exit plane (all z and r at $\theta = X_H = +10$)

$$\frac{\partial V_\theta}{\partial \theta} = \frac{\partial V_r}{\partial \theta} = \frac{\partial V_z}{\partial \theta} = 0 \text{ with overall continuity imposed.}$$

Thus no *a priori* assumption was made regarding the presence or otherwise of recirculation within the bend and the upstream and downstream boundary conditions were assigned in the straight lengths of duct upstream and downstream of the bend where the flow may reasonably have been assumed to be parabolic.

The time required for a converged solution with $10 \times 15 \times 60$ nodes was around 80 min of CDC6600 with a convergence criterion of a maximum residual, in any equation, of 10^{-3}.

10.16 Square-duct bend, 2

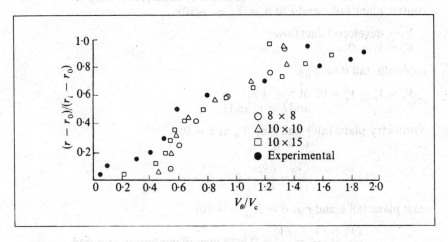

The use of $10 \times 15 \times 20$ grid nodes in each of the three parts of the solution was a compromise. The time required for converged solutions was $(I \times E \times S \times N) \, 2 \times 10^{-4}$ s for E equations, S sweeps of the algorithm and I iterations on a grid with N nodes. Subsequent runs on similar geometries, using the pressure field from the present calculations as a first approximation allow, as might be expected, a converged solution to be achieved with many fewer iterations.

The number of grid nodes was selected after extensive numerical testing. As indicated by Humphrey et al., (1977) the solution on the line of symmetry and at 90 degrees changes significantly with grid nodes ranging from 8×8 to 10×10 to 10×15 grid nodes in a cross-stream plane and those obtained with the finest grid were still different from experimental results which were believed to be of high quality. It was clear that the 10×15 grid mesh resulted in smoothing out of velocity peaks in the flow but the computer time and cost requirements made finer grids unacceptable.

The panel indicates the result of the test calculations and the influence of the number of nodes is obvious.

The calculated results of Humphrey et al. (1977) indicated the existence of a region of recirculating flow between the entrance to the bend and 18 degrees downstream and close to the outer radius. Corresponding measurements, with dye traces, confirmed this region of recirculation and demonstrated that the traces curved from the centre of the duct, laterally, before turning against the direction of the main flow within the 10 per cent of the duct close to the side walls. The solution of the full three-dimensional, elliptic equations was necessary to represent this feature of the flow.

10.17 Square-duct flow bend, 3

Solutions can be obtained with fully elliptic equations of with reduced forms corresponding to

$\dfrac{\partial^2}{\partial \theta^2} = 0$ partially parabolic

$\dfrac{\partial^2}{\partial \theta^2} = 0$ and $\dfrac{\partial P}{\partial \theta} = f(\theta)$ with f prescribed   parabolic

$v = 0$ and finite vorticity   ideal, rotational

$v = 0$ and zero vorticity   ideal, irrotational or potential

It is important to ask what reduction in precision may be associated with the solution of reduced forms of equations.

The solution of potential-flow equations is comparatively simple and inexpensive. It cannot, of course, lead to values of skin friction drag or predict secondary flows but, as shown for example by Ward-Smith (1971), useful calculations of pressure coefficient can be obtained. Rotational-flow equations have also been shown, for example, by Squire and Winter (1951) to represent flows with curvature provided the bend angle and radius ratio are small and the aspect ratio large. The secondary flows, for example, have the correct trends provided the side walls are far apart; the quantitative discrepancies can, however, be considerable even for a large aspect ratio situation. Again, the influence of skin friction drag is not represented and the pressure distributions are similar to those with irrotational flow. In turbulent bend flow, in contrast, potential flow does represent pressure distributions of bends with aspect ratios equal to 1 or larger with precision satifactory for engineering purposes, provided the radius ratio is not too severe (see Ward-Smith, 1971).

The calculations of Humphrey et al. (1977) indicated that in the 45° plane, the longitudinal diffusion was usually less than 0·05 of the total diffusion; in addition, it was usually less than 0·001 of the longitudinal convection. Thus, the neglect of longitudinal diffusion should lead to approximately correct results for substantial regions of the flow. The representation of the flow by parabolic equations is, however, unlikely to lead to reasonable results in view of the substantial radial variations in longitudinal pressure gradient observed by Humphrey et al. (1977). The use of the so-called "partially parabolic" equations is more likely to be successful although they cannot represent regions of recirculating flow and these may be significant.

The main difficulty inherent in the use of reduced-form equations is the need for *a priori* information that there is no region of separated flow or significant radial pressure variation.

10.18 Square-duct flow bend, 4

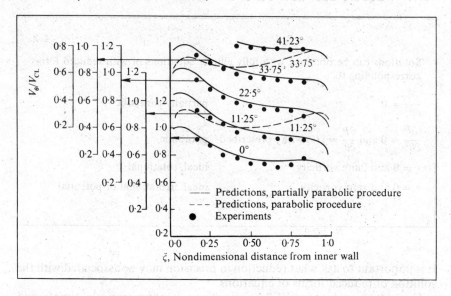

The above results are reproduced from the paper by Pratap and Spalding (1975) and correspond to their bend which has a mean radius of curvature of 2·52 m and an aspect ratio of approximately 4 with the larger dimension in the radial direction. The duct was preceded by a straight section of 1·22 m and the flow was turbulent.

In this case, the curvature is not particularly strong and the partially parabolic scheme might be expected to correspond reasonably with experiments. As the panel shows, this suggestion is supported by a comparison with the experiments of Howard *et al.* (1975) at a location fairly close to the duct centreline at various downstream positions. Values of longitudinal velocity calculated with the parabolic procedure are, however, shown to be different from those with the partially parabolic procedure although, for this curvature, this might not have been anticipated. The two-equation turbulence model used for the calculation takes no account of normal stress-driven secondary flows and, although these are likely to be small compared to the pressure-driven secondary flows, the turbulence model can undoubtedly be improved. It is, therefore, the difference between the results obtained with the two numerical procedures which is of greatest importance: it remains to be demonstrated whether the results would be different again if a fully elliptic scheme were used.

In their partially parabolic procedure, Pratap and Spalding (1975) reported the use of a 17 × 14 grid mesh in the cross-stream direction with 10 stations in their upstream straight section and 24 in the curved section with an approximate spacing of 5 degrees. The partially parabolic procedure converged in

Recirculating Flows

50 sweeps with a computing time of 15 minutes. The corresponding computing time for the parabolic case was around 5 minutes. These times may be compared directly with those for the elliptic equations if allowance is made for the turbulence equations solved by Pratap and Spalding. As a result, it is clear that there are significant differences in the computing requirements for the fully elliptic partially parabolic and parabolic equations. The absolute time values should not be taken too literally in view of possible differences in initial guesses in the flow conditions themselves and in the optimization of the computer programs.

Measurements with turbulent flow in the bend of panel 10.15 have been reported by Humphrey and Whitelaw (1977) and have been considerably extended by Humphrey, Whitelaw and Yee (1980) and by Taylor, Whitelaw and Yianneskis (1980). Related calculations, with a two-equation turbulence model and the numerical solution procedure of Humphrey *et al.* (1977) have been reported by Humphrey *et al.* (1980) and suggest that the precision of calculations is limited more by the numerical assumptions than by the turbulence model.

10.19 Jet in a crossflow

- As for the flow situations of the two previous panels, a numerical procedure can readily be developed to solve three-dimensional elliptic equations
- The run cost and storage requirements were comparatively large and limit the attainable precision probably to a greater extent than turbulence models
- The general characteristics of jets in crossflows have been adequately represented although numerical smearing of velocity peaks can exist

Calculations of a jet in a crossflow require the solution of three-dimensional, elliptic equations and have been reported, for example, by Patankar et al. (1977), Jones and McGuirk (1979a) and Crabb (1979). Previous measurements have shown that the jet bends and takes on a horseshoe shape (in planes orthogonal to the flow direction) which persists with downstream distance, and it can be expected that this structure is pressure controlled with comparatively small differences between laminar- and turbulent-flow results. Thus, the influence of the turbulence model should be small. All three sets of calculations made use of two-equation models and numerical schemes based on pressure-corrector algorithms. The results of Crabb, for example, used $22 \times 15 \times 15$ node points in the longitudinal, transverse and cross-stream directions, respectively: 50 000 words of CDC6600 store were required with run times of the order of 25 min.

The results of the three investigations show significant differences which are most likely due to boundary condition or numerical assumptions. In the case of Patankar and Spalding (1972) the agreement with experiment was probably enhanced by the location of the numerical boundaries and, in particular, the surface above the jet: in the same way as in an experiment. moving the top surface towards the jet exit squeezes the jet trajectory and, in this case, may have overcome other calculation errors. In addition, their comparison with the measurements of Ramsey and Goldstein (1972) is misleading in that hot-wire measurements give incorrect answers in the vicinity of regions of flow recirculation. The results of Jones and McGuirk are in reasonable agreement with measurements although those of Crabb, with similar numerical and turbulence assumptions, are less so and tend to smear velocity peaks. The most likely reason for the difference lies in the chosen distributions of grid nodes which is particularly important in the region of maximum jet curvature. The cost of computations has so far precluded a satisfactory investigation of the influence of grid distribution.

For small velocity ratios, reduced equations similar to those discussed in connection with bend flows, can be appropriate. Bergeles *et al.*, (1978) have described solutions of appropriate reduced equations for low jet velocities.

Note added in proof.

The recent results of A. J. White (ASME Paper 80-WA/HT-26) show the dependence of jet-in-crossflow calculations for node distributions up to $25 \times 20 \times 15$; it is concluded that more nodes are required to achieve grid independence.

10.20 Concluding remarks

Three main conclusions may be extracted from the preceding text, i.e.

- Elliptic equations for two-dimensional recirculating flows, can be solved with sufficient precision for many engineering applications.
- Experimental evidence does not suggest large benefits from the use of existing turbulence models with more than two equations; improved turbulence models are very desirable and are likely to require changes to the assumptions for the pressure strain term and for the dissipation equation.
- Elliptic equations for three-dimensional recirculating flows, can be solved. In some cases, adequate precision cannot be achieved due to computer storage limitations and finite-difference assumptions.

They may usefully be contrasted with the situation for boundary-layer flows where more than adequate numerical precision can usually be obtained. In boundary-layer flows, transport models can be less important and many wall boundary-layer flows can be calculated with algebraic eddy viscosity hypotheses such as that of Cebeci and Smith (panel 3.6). Computer storage and cost are less important in boundary-layer flows and seldom limit the precision of calculations. In recirculating flows, numerical limitations still exist and, particularly in flows controlled more by pressure forces than turbulence, can limit the precision of results. The present turbulence models are undoubtedly in error when applied to bluff-body stabilized flows and to highly swirling flows and improved assumptions are required.

The third conclusion may become less important with the arrival of larger and faster computers but it is more likely that numerical experiments and consequent advances in our understanding of the relative advantages of different techniques for different classes of flow will lead to the desired improvements.

11 Viscous–Inviscid Interactions and Corner Flows

11.1 Purposes and outline of chapter

- To discuss interaction between a thin shear layer and an external inviscid flow, using the displacement-thickness concept to match solutions for the two flow regions and thus avoid solving the complete Navier–Stokes equations

- To discuss the extension to cases where a Navier–Stokes solution is needed for the shear layer and the displacement-thickness concept is not useable.

- To discuss the extra difficulties of slender shear layers, particularly flows along streamwise corners

This chapter reviews two cases in which thin shear-layer concepts need modification; one is mainly a numerical problem while the other involves changes in turbulence structure as well.

The "matched asymptotic expansion" view of boundary layer theory (Van Dyke, 1975) implies that $\partial U/\partial y$ in the free stream is negligible compared to that in the shear layer (this in turn implies $\partial P/\partial y \ll \rho U_e^2/\delta$). If the free stream velocity is truly independent of y all external-flow streamlines are displaced the same distance by the presence of the shear layer; this distance is the displacement thickness δ^*. On highly curved surfaces and in the important case of flow near an airfoil trailing edge, $\partial U_e/\partial y$ and $\partial P/\partial y$ are not negligible, so the thin shear-layer equations are inadequate and must be replaced by a Navier–Stokes solution in the shear layer. It is shown that this need not involve large increases in computing time.

In "slender" shear flows such as jets from circular or rectangular nozzles, wakes of submarines or missiles, and flow in ducts or other streamwise corner flows, large gradients of streamwise velocity U appear in both cross-stream coordinate directions (y and z). Use of cylindrical polar coordinates in axisymmetric flows leaves only one large gradient (in the radial direction) but the complications of turbulence behaviour in slender flows are hidden, rather than absent, in axisymmetric flow. In nonaxisymmetric cases secondary flow (streamwise vorticity) can be generated by Reynolds stress gradients in the yz plane; Reynolds stress gradients will generally attenuate a secondary flow generated by inviscid mechanisms.

11.2 Basic concepts of interactions

> - For thin shear layer calculations U_e or P_e must known but, unless δ^* is very small, the presence of the shear layer affects the outer flow via "displacement effect"
>
> - An alternative to solving the Navier–Stokes equations for the whole flow field is to match the boundary-layer calculation to "inviscid" flow calculation by iteration
>
> - The lifting airfoil is a difficult case because of the presence of strong upstream influence of the flow near the trailing edge

The boundary conditions for thin shear layers include specifications of the mean velocity (and hence the pressure) at the edge or edges. In any real flow this velocity is determined by the behaviour of the "inviscid" flow external to the shear layer. Strictly it is the stress gradients rather than the viscosity that become negligible but the brief, imprecise label is commonly used. If the shear layer is thin, the external flow *may* be almost the same as if the shear layers were absent and the flow inviscid everywhere, but there are some cases, such as a lifting airfoil, where the effect of even a thin shear layer on the pressure distribution is large.

The pressure distribution in duct flow is also significantly affected by shear layers whose thickness δ is not very small compared to the duct height h, even if the shear layer is "thin" in the sense that $d\delta/dx$ is very small.

If the division into shear layer and inviscid flow is to be kept, the shear-layer calculation and the inviscid calculation must be repeated alternately until they match. In duct flows with small curvature the inviscid calculation and the matching process are simple, since upstream influence is negligible (panel 5.2). The shear-layer calculation is also relatively simple, since the thin-shear-layer equations apply. External flows are more difficult.

11.3 Displacement effects on an airfoil

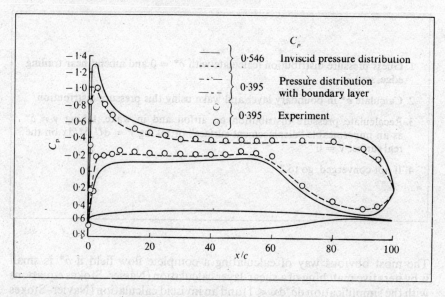

This panel shows how viscous effects can affect the lift coefficient of an airfoil even at its design incidence, in the absence of any spectacular phenomena like shock-induced separation or stall. The displacement thickness, $\delta^* = \int_0^\delta (1 - \rho U/\rho_e U_e)\,dy$, changes the effective shape seen by the inviscid external flow so that

i. the apparent camber is decreased because the upper-surface boundary layer is thicker than the lower-surface one near the trailing edge;
ii. the pressure drag becomes nonzero because the pressure at the trailing edge no longer rises to the stagnation value; the pressure distribution near the trailing edge depends on the displacement thickness in the near-wake as well as on the airfoil, and differs greatly from the pressure distribution in inviscid flow where C_p tends to unity at the trailing edge.

If the streamlines are highly curved, e.g. near the trailing edge, the pressure difference across the shear layer will be significant, the above definition of displacement thickness is not sufficient, and the mean velocity distributions either side of the trailing edge will be different, implying that some circulation is carried by the wake.

Note that the changes in surface pressure caused by the shear layer are usually far larger than the pressure difference across the shear layer: there is no close connection between the two.

11.4 Iterative calculation of δ^*-interaction

---

1. Guess pressure distribution (calculate with $\delta^* = 0$ and amend near trailing edge, say)
2. Calculate δ^* in boundary layer and wake using this pressure distribution
3. Recalculate pressure distribution on airfoil and in wake, taking $y = \delta^*$ as an impermeable "displacement surface" or taking $V = \mathrm{d}U_e \delta^*/\mathrm{d}x$ on the real surface $y = 0$
4. If not converged, go to 2

The most obvious way of calculating a complete flow field if δ^* is small is by iterative matching of a shear-layer calculation (Navier–Stokes equations with the simplification $\mathrm{d}\delta/\mathrm{d}x \ll 1$) and an inviscid calculation (Navier–Stokes equations with the simplification "stress gradients $\ll$ pressure gradients") for the external flow.

The iteration will converge only if the change in δ^* caused by a small change in pressure distribution, is small everywhere; it is bound to diverge near the stall for this reason, irrespective of whether the thin shear layer equations are still accurate.

Strictly (see panel 11.9) the inviscid flow calculation should be performed only for $y > \delta$, the boundary condition at $y = \delta$ being the value of V obtained from the shear-layer calculation. If the thin shear layer approximation $\mathrm{d}\delta/\mathrm{d}x \ll 1$ holds, it is accurate enough to perform the inviscid calculations for all $y > 0$, the boundary condition on $y = 0$ being $V = \mathrm{d}U_e \delta^*/\mathrm{d}x$. Equivalently we can specify zero normal velocity on the line $y = \delta^*$ (the "displacement surface" or "fluid airfoil").

For a fuller discussion see Cebeci and Bradshaw (1977, Chapter 11).

11.5 Best use of surface-singularity methods

1. Choose body shape and invert "influence-coefficient" matrix
2. Replace $d\delta^*/dx$ by an injection velocity ratio V_w/U_e
3. Evaluate tangential velocity distribution by adding velocity field of injection sources to solution for solid body
4. Repeat 2 and 3 for new δ^*, changing only the injection part of the solution

In surface singularity methods for inviscid-flow calculations, such as the Douglas–Neumann programs (Hess and Smith, 1967), much of the computation time is taken up by inverting the geometry-dependent matrix relating induced velocity at each point on the body to source density at all other points. Once the matrix has been inverted, calculation of the tangential velocity at the body surface is rapid. The change from a solid surface to one with specified transpiration velocity is trivial.

Note that these methods can be used for internal flows by specifying the normal velocities at the inlet and outlet planes. Errors occurring near the corners can be compensated by near-universal fractional changes to the normal velocities of the corner elements.

In compressible flow, relaxation methods must be used. Usually the flow outside a two-dimensional body is transformed to the flow inside a circle. Again it is possible to make the geometry-dependent transformation once and for all, and change the injection velocity at each iteration. Typically several relaxation iterations are needed between each recalculation of the boundary layer and wake.

11.6 Upstream influence

- Subsonic flows admit upstream influence (disturbance at $x = x_0$ affecting flow at $x < x_0$ (see panel 5.2)
- Thin shear-layer equations forbid upstream influence
- In subsonic flow enough upstream influence may propagate through the external flow
- In supersonic external flows any upstream influence must be allowed to propagate through the subsonic part of the shear layer

The Navier–Stokes equations or the "inviscid" Euler equations are elliptic in subsonic flow (the latter but *not* the former being hyperbolic in supersonic flow). This means that the effects of a disturbance can propagate upstream. As stated in Chapter 5 the boundary-layer equations are parabolic and the influence of a disturbance at a point (x_0, y_0) is confined to the half-plane $x > x_0$: paradoxically it is mainly the neglect of $\partial P/\partial y$ that restricts x-wise propagation. This means that difficulties may occur when the boundary-layer equations are used in a problem where upstream influence within the shear layer is significant. In most subsonic external flows, where a boundary-layer calculation is matched to an inviscid external calculation, the upstream influence carried by the inviscid flow is enough to avoid numerical breakdown and probably to give acceptably accurate results. In the case of supersonic external flows any upstream influence must propagate through the subsonic part of the shear layer and the boundary-layer equations cannot be used in the normal way. The alternatives are to use the full (time-averaged) Navier–Stokes equations in the shear layer, or to put the upstream influence in by adjustment of the initial conditions of a boundary-layer calculation as in the Crocco–Lees methods (e.g. Alber and Lees, 1968). A good review with special reference to shock boundary-layer interaction is given by Green (1970): see also Roache (1972b). The reader may also like to review the material on classification of partial-differential equations in Chapter 5.

In subsonic flow, upstream influence within the shear layer is important only in a few cases, such as the neighbourhood of an airfoil trailing edge or region of large streamline curvature.

11.7 Shear layers with $\partial P/\partial y \neq 0$

---

1. Guess $P(x, y)$ within the shear layer [$P(x, \delta)$ being a boundary condition]
2. Calculate U, V and $\overline{u_i u_j}$ from the x-momentum equation, continuity equation and stress equations
3. After each x step of 2, recalculate $\partial P/\partial y$ from the y-momentum equation. Deduce $P(y)$ and store
4. After the sweep in x is finished, go to 2 using the new $P(x, y)$. Repeat until converged

---

The Reynolds averaged Navier–Stokes equations can be solved by successive approximation in any fairly thin shear layer. They are elliptic, and upstream influence within the shear layer is simulated numerically by repeated sweeps of a conventional marching calculation over the full range of x: upstream influence progresses at one x-step per iteration. The key to the process is that step 2 uses the $P(x, y)$ field obtained from the previous sweep: newly calculated values are not used, because sweep-to-sweep changes would upset finite-difference values of $\partial P/\partial x$. Only $P(x, y)$, not U, is stored. This is sometimes called a "partially parabolic" procedure, as indicated in the preceding chapter, but the *equations* are fully elliptic.

Mahgoub and Bradshaw (1979) have added a $\partial P/\partial y$ calculation to the basic boundary-layer calculation method of Bradshaw and Unsworth (1977). At present, normal stresses are related to shear stress by algebraic formulas but full transport equations could be added.

This is a use of the general argument that the terms which are neglected in the thin shear-layer equations will almost always be small enough to *approximate* (panel 1.6). Approximations of the smaller turbulent stress terms can equally be regarded as part of the turbulence model. It may be noted that conversion to (s, n) coordinates is necessary and is much easier if the equations are all of first order. An important application is to curved shear layers and the flow near the trailing edge of an airfoil.

11.8 Shear layer boundary conditions with $\partial P/\partial y = 0$

1. Boundary layer

   $y = 0$; $U = 0$, $V = V_w$ or 0, P_w not given
   $y = \delta_0 \geqslant \delta$; U_e or P_e given, V_e not given
   — find V_e (and P_w)

2. Wake

   $y \leqslant \delta_l$; U_l or P_l given

   $y \geqslant \delta_u$; U_u or P_u given
   — find V_l, V_u

In external flow the real-life problem is to match the shear layer to the external "inviscid" flow at $y = \delta$. Therefore for a boundary layer the boundary conditions are the conventional ones except that U_e or P is specified at $y = \delta_0$, an arbitrary position just outside $y = \delta$. Strictly, surface-singularity methods yield U_e, given V_e, and P_e follows from Bernoulli's equation: we can no longer assume V_e/U_e very small although V_e^2/U_e^2 will be very small in most cases. Most calculation methods can accept P_w as input and, although this does not correspond to any real problem, it may be useful for comparisons with experiments.

In a wake, the boundary conditions available from the inviscid flow calculations are the pressures P_u and P_l at the upper and lower edges, given the V component velocities at these edges. Thus the shear layer problem has P_u and P_l specified, V_u and V_l unspecified. V_l must be chosen, as a function of x, to yield the right pressure difference across the wake—in effect, we have to calculate the curvature of the wake. At any given x step the exact algebraic equation for V_l will be slightly nonlinear because of the $V\partial V/\partial y$ term in the y-component momentum equation. Once V_l has been chosen the calculation is equivalent to that for the boundary layer (in practice the main calculation and the evaluation of V_l will be combined).

11.9 Interaction calculations with $\partial P/\partial y \neq 0$

1. Guess $\delta_0(x)$ and $V(y = \delta_0)$—not critical: $\delta_0 \not< \delta$.
2. Use surface-source method to calculate flow about the surface $y = \delta_0$ with given V
3a. Calculate shear layer, given P at $y = \delta_0$, by one sweep of the procedure in the previous panel
3b. On the first cycle only, reset δ_0 to 1·2 times the calculated δ, (say)
4. Return to 2, using V from the shear layer calculation

This is the procedure used by Mahgoub and Bradshaw (1979). It does not use the displacement thickness (panel 11.3) but matches the inviscid-flow calculation to the shear-layer calculation at $y = \delta_0$; V from the shear-layer calculation is the boundary condition for the inviscid calculation, P from the inviscid calculation is the boundary condition for the shear-layer calculation. Note that a shear-layer calculation will work perfectly well just outside the shear-layer edge, although it is rather uneconomical. Therefore the choice of δ_0 is not critical, and for most purposes can be reset, generously, only on the first cycle. Thus the "body" shape for the inviscid calculation can be kept the same for all succeeding cycles (see panel 11.5).

A similar procedure has been used, with an integral shear-layer calculation, by Patel and Nakayama (1976).

11.10 Extra difficulties of slender shear layers

---

1. $\partial U/\partial z \sim \partial U/\partial y$ (but $\gg \partial U/\partial x$) in contrast to the thin shear-layer where $\partial U/\partial z \sim \partial U/\partial x \ll \partial U/\partial y$
2. V and W are influenced by normal-stress gradients
3. Opposing pressure gradients in yz plane are important
4. Streamwise vorticity $\Omega_x \equiv \partial W/\partial y - \partial V/\partial z$ is significant and $\partial V/\partial z \sim \partial W/\partial y$

In slender shear layers elongated in the x direction (e.g. streamwise corner flows, or near-axisymmetric wakes or jets), $\partial/\partial x \ll \partial/\partial y \sim \partial/\partial z$. The only permissible simplification of the Navier–Stokes equations is the neglect of x-wise gradients of viscous stress or Reynolds stress. The V and W motions —the "secondary flow"—are more complicated than in three-dimensional thin shear layers: in a three-dimensional boundary layer, the x-component vorticity is $\partial W/\partial y$, while in a slender shear flow it is $\partial W/\partial y - \partial V/\partial z$, the second term being of the same order as the first.

Significant problems arise in turbulence modelling and in numerical solutions (notably in the boundary conditions).

11.11 Two types of secondary flow

1. Skew-induced (Prandtl's first kind)

$$\frac{D\Omega_x}{Dt} = \Omega_x \frac{\partial U}{\partial x} + \Omega_y \frac{\partial U}{\partial y} + \Omega_z \frac{\partial U}{\partial z} + \text{(below)}$$

— Vortex lines

2. Stress-induced (Prandtl's second kind)

$$\frac{D\Omega_x}{Dt} = \text{(above)} + \left(\frac{\partial^2}{\partial y^2} - \frac{\partial^2}{\partial z^2}\right)(-\overline{vw}) + \frac{\partial^2}{\partial y \partial z}(\overline{v^2} - \overline{w^2}) + \nu\nabla^2\Omega_x$$

The first three terms on the right of the streamwise vorticity equation are the intensification of existing Ω_x by stretching, which could be important in accelerated or retarded flows in diffusers, and the augmentation of Ω_x through skewing of existing Ω_y and Ω_z by the mean shear. In an initially two-dimensional boundary layer, say on an aeroplane body with y spanwise, $\Omega_z = -\partial = -\partial U/\partial y$ is the only large component of vorticity. Large values of $\partial U/\partial z$ (z vertical) occur in the mean pressure field near the leading edge and the z-wise vortex lines are wrapped round the wing with strong trailing vorticity imbedded in the slender shear layer in the wing-body junction. Neither turbulence nor viscous stresses are necessary except to generate the boundary layer in the first place. Lift merely makes the flow asymmetrical.

The $\nu\nabla^2\Omega_x$ term represents viscous diffusion of existing vorticity. The Reynolds-stress terms can either increase or decrease vorticity in a given region of the flow, though $\iint \Omega_x \, dy \, dz$ is unaltered (as required by conservation of angular momentum). In a symmetrical streamwise corner a vortex pair arises: see panel 11.14.

Note that P does not appear in the Ω_x equation. Pressure gradients do appear in its boundary conditions, and the ratio of the two constituents of Ω_x, $\partial W/\partial y$ and $\partial V/\partial z$, depends on $\partial^2 P/\partial y \partial z$.

11.12 Decay of skew-induced vorticity

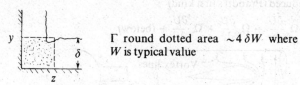

Γ round dotted area $\sim 4\delta W$ where W is typical value

$\overline{vw}$ is probably of order $\overline{uv}\, W/U \sim 0\cdot 002\, UW$. Thus

$D\Gamma/Dt \approx -2\overline{vw} \sim -0\cdot 001\, \Gamma U/\delta$

so time constant $\sim 1000\, \delta/U$; distance travelled $\sim 1000\, \delta$. Actually δ grows!

A very crude calculation of the decay of the longitudinal circulation in a corner due to the y and z components of surface shear stress can be made as shown. W is just a typical value of V or W velocity in the corner. The assumption that $\overline{vw}/\overline{uv} \simeq W/U$ is likely to be closer than assuming proportionality to $(W/U)^2$.

For practical purposes Γ is *conserved*! Thus W will decrease proportional to $1/\delta$ as δ grows, and Ω_x, which is of order W/δ, will vary as $1/\delta^2$.

Inequality of the boundary layer thicknesses on the wing and the body will complicate the picture. The fractional variation of body δ from leading edge to trailing edge is, however, not all that large.

If the same arguments are applied to a vortex imbedded in a shear layer far from a corner (either in a nearly plane boundary layer or in a free shear layer) they again suggest that W/U decreases little faster than $1/\delta$.

11.13 Wing-body junctions

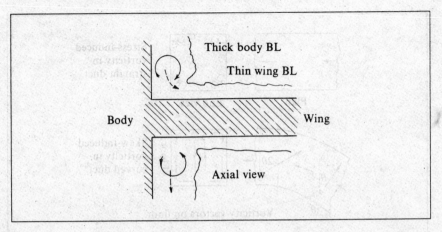

A wing-body (or blade-hub) junction is completely dominated by skew-induced secondary flow arising at the leading edge. Except in the case of a very sharp leading edge the horseshoe vortex will be strong enough to persist all the way to the trailing edge. Reynolds stress (and viscous stress) gradients in the junction will generally be in such a direction as to *weaken* the streamwise vortex. The double-vortex pattern of stress-induced secondary flow would appear only after the skew-induced vortex had been greatly attenuated —probably far downstream of the trailing edge.

If the wing lift coefficient is high the vortex may move away from the corner, the upper (suction-surface) vortex drifting out on to the wing and the pressure-surface vortex drifting down on the body as shown dotted above. In a cascade the latter vortex can reach the suction surface of the blade below (Chapter 13).

11.14 Vorticity in a square duct

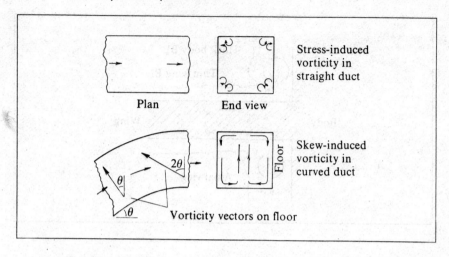

Many experiments on stress-induced vorticity in ducts have been reported (see Melling and Whitelaw, 1976). Long straight polygonal ducts are quite common in engineering practice, but any longitudinal curvature results in skew-induced vorticity: to a first inviscid-flow approximation, a right-hand bend of θ degrees skews the vortex lines in the floor boundary layer θ degrees to the left—so the angle between primary streamlines and vortex lines, initially 90°, is now $90 + 2\theta°$. In axes aligned with the original flow direction, y vertical, z lateral, $W/U \approx -(\partial W/\partial y)/(\partial U/\partial y) = -\Omega_x/\Omega_z$. The flow pattern is quite different from the stress-induced pattern, and the turbulence-modelling problems are so different that one cannot be sure that the same calculation method can be used in both configurations (see Chapter 10).

The magnitude of secondary flows caused by the imbalance between the centrifugal and pressure-gradient terms increases rapidly with curvature and leads to secondary velocities which can be more than 30 per cent of the longitudinal mean velocity in ducts of practical curvature. In contrast, normal-stress-driven secondary velocities are usually of the order of 1 per cent of the longitudinal mean velocity. The former, being pressure driven, do not necessarily require a complex turbulence model; in the latter, the normal stresses must be considered as discussed, for example, by Tatchell (1975) and Gessner and Emery (1979).

11.15 Streamwise vorticity in free shear layers

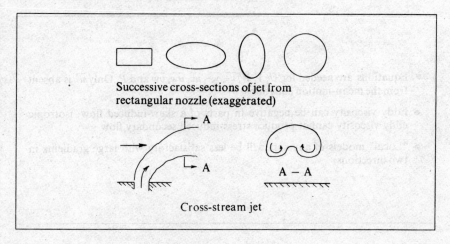

Successive cross-sections of jet from rectangular nozzle (exaggerated)

Cross-stream jet

Noncircular jets or wakes are subject to stress-induced secondary flows and the approach to a circular cross-section far downstream may be non-monotonic (Sforza et al., 1966). Cross-stream jets or plumes, and the wakes of lifting bodies, suffer from skew-induced secondary flows (plus or minus the effects of lift-induced trailing vortices).

Classical wingtip-type or delta wing-type trailing vortices form from the rolling up of a three-dimensional mixing layer. When fully rolled up, the central core is strongly stabilized by the outward increase of angular momentum about the vortex axis; it is effectively laminar, and disturbances can propagate for long streamwise distances without attenuation. Similar stabilization may be found in other types of strong vortex (Bradshaw, 1973).

For numerical methods for the initial region of a cross-stream jet (with strong upstream influence) see panel 10.19. There have been results, for example in McGuirk and Rodi (1979) who used a three-dimensional form of the Navier–Stokes equations to represent the data of Sforza et al. (1966) and Sfeier (1975). The calculations appear to be successful with a relatively simple turbulence model although discrepancies occur in the upstream or downstream regions, depending on the turbulence-model constants.

11.16 Turbulence modelling in slender flows

- Equations are needed for U, V, W, $\overline{v^2}$, $\overline{w^2}$, $\overline{uv}$, $\overline{uw}$, $\overline{vw}$ and P. Only $\overline{u^2}$ is absent from the mean-motion equations
- Eddy viscosity can be negative in parts of a skew-induced flow; isotropic eddy viscosity cannot produce stress-induced secondary flow
- "Local" models of any sort will be less satisfactory with large gradients in two directions

---

Except for rough-and-ready purposes, nothing short of a full Reynolds stress model can accurately simulate the growth of stress-induced secondary flow or the decay of skew-induced secondary flow. An isotropic eddy viscosity cannot simulate stress-induced secondary flow at all (this kind of secondary flow does not occur in laminar flow) and recent measurements (Shabaka, 1979) have shown that $-\overline{uv}$ is large and of opposite sign to $\partial U/\partial y$ over most of the "wing" side of a wing-body junction with y spanwise. The latter phenomenon is also found in stress-induced secondary flows but there $-\overline{uv}$ is small in the region of negative eddy viscosity.

One of the main points of difficulty in models for two-dimensional boundary layers is the effect of the solid surface on the pressure–strain term and the length scale. In a corner flow we have two solid surfaces.

Considerable efforts have been made to model stress-induced secondary flows although in most engineering applications they are relatively unimportant. These include the contributions of Launder and Ying (1973), Tatchell (1975), Hanjalić and Launder (1972) and Gessner and Emery (1979).

12 Stability and Transition

12.1 Purposes and outline of chapter

- To discuss the derivation of the Orr–Sommerfeld equation for two- and three-dimensional flows
- To discuss the numerical solution of the Orr–Sommerfeld equation and the computation of the eigenvalues
- To present stability curves for some two- and three-dimensional flows
- To discuss the prediction of transition in two- and three-dimensional flows by the e^9-method

In this chapter we first discuss the derivation of the stability equation, known as the Orr–Sommerfeld equation, for two- and three-dimensional flows with the parallel-flow approximation and by assuming that the disturbance is a sinusoidal travelling two-dimensional wave. This is followed by a discussion concerning the procedure for computing the eigenvalues of the Orr–Sommerfeld equation for two- and three-dimensional problems using spatial amplification theory. The numerical procedure which we use to solve this equation is the Box method. The computation of the eigenvalues and eigenfunctions is described in detail and results are presented for both two- and three-dimensional flows.

A principal reason for solving the Orr–Sommerfeld equation relates to its possible application to prediction of the transition from laminar to turbulent boundary-layer flow. So far we have represented transition by equations which are entirely empirical and, as we shall see, the Orr–Sommerfeld equation and the e^9-method offer the possibility of calculating transition with great generality and much reduced empiricism.

12.2 Parallel flow approximation and small disturbance equations

> Assume $U(y)$, $V(y) = 0$, $W(y)$ (Parallel Flow Approximation) and add fluctuating parts u, v, w and p to the mean values of $U, W,$ and P and substitute them in equations (5) to (8). Retaining only the linear terms in u, v, w, we obtain
>
> $$\frac{\partial u}{\partial T} + U\frac{\partial u}{\partial X} + v\frac{\partial U}{\partial Y} + W\frac{\partial u}{\partial Z} = -\frac{\partial p}{\partial X} + \frac{1}{R}\nabla^2 u \quad (1)$$
>
> $$\frac{\partial v}{\partial T} + U\frac{\partial v}{\partial X} + W\frac{\partial v}{\partial Z} = -\frac{\partial p}{\partial Y} + \frac{1}{R}\nabla^2 v \quad (2)$$
>
> $$\frac{\partial w}{\partial T} + U\frac{\partial w}{\partial X} + v\frac{\partial W}{\partial Y} + W\frac{\partial w}{\partial Z} = -\frac{\partial p}{\partial Z} + \frac{1}{R}\nabla^2 w \quad (3)$$
>
> $$\frac{\partial u}{\partial X} + \frac{\partial v}{\partial Y} + \frac{\partial w}{\partial Z} = 0 \quad (4)$$

With the introduction of the dimensionless quantities

$$X = \frac{x}{l},\ Y = \frac{y}{l},\ U = \frac{\hat{U}}{U_0},\ P = \frac{\hat{P}}{\rho U_0^2},\ V = \frac{\hat{V}}{U_0},\ W = \frac{\hat{W}}{U_0},\ T = \frac{tU_0}{l},\ R = \frac{U_0 l}{\nu}$$

where l is a characteristic length of the flow (e.g. the displacement thickness). The Navier–Stokes equations can then be written as

$$\frac{DU}{DT} = -\frac{\partial P}{\partial X} + \frac{1}{R}\nabla^2 U, \quad (5)$$

$$\frac{DV}{DT} = -\frac{\partial P}{\partial Y} + \frac{1}{R}\nabla^2 V, \quad (6)$$

$$\frac{DW}{DT} = -\frac{\partial P}{\partial Z} + \frac{1}{R}\nabla^2 W, \quad (7)$$

$$\frac{\partial U}{\partial X} + \frac{\partial V}{\partial Y} + \frac{\partial W}{\partial Z} = 0. \quad (8)$$

Here

$$\frac{D}{DT} = \frac{\partial}{\partial T} + U\frac{\partial}{\partial X} + V\frac{\partial}{\partial Y} + W\frac{\partial}{\partial Z}. \quad (9)$$

The procedure of the panel is now followed.

Here u, v, w and p are small enough for second-order products like $\overline{uv}, \overline{wv}$ to be negligible. The mean velocity satisfies the three-dimensional, thin-shear-layer equations for laminar flows.

Equations (1) to (4) are subject to the usual boundary conditions,

$$u(0) = v(0) = w(0) = 0 \text{ at the wall}$$

and disturbances go to zero as $Y \to \infty$.

12.3 Modal analysis or separation of variables

Define complex amplitude functions of the flow variables by

$$(\hat{u}, \hat{v}, \hat{w}, \hat{p}) = (\hat{f}, \hat{\phi}, \hat{h}, \hat{\Pi}) \exp[i(\hat{\alpha}x + \hat{\beta}z - \hat{\omega}t)] \qquad (1a)$$

or, in terms of dimensionless quantities

$$(u, v, w, p) = (f, \phi, h, \Pi) \exp[i(\alpha X + \beta Z - \omega T)] \qquad (1b)$$

Here

$$\alpha = \hat{\alpha}l, \qquad \beta = \hat{\beta}l$$

$$\omega = \frac{\hat{\omega}l}{U_0} \qquad (2)$$

We assume that the small disturbance is a sinusoidal travelling plane wave and write a three-dimensional disturbance as shown in equation (1). Here f, ϕ, h, Π, which are functions of y only, are the complex amplitude functions of u, v, w and p, respectively; α, β and ω are complex quantities in general. The wavelengths of the disturbances in the X- and Z-directions are respectively $\lambda_x = 2\pi/\alpha_r$, $\lambda_z = 2\pi/\beta_r$, and the radian (circular) frequency of the disturbance is ω_r.

12.4 Modal equations

> Using the definition of equation (1b) of panel 12.3, the equations (1) to (4) of panel 12.2 become
>
> $$f'' - \xi^2 f - RU'\phi - iR\alpha\Pi = 0 \qquad (1)$$
> $$h'' - \xi^2 h - RW'\phi - iR\beta\Pi = 0 \qquad (2)$$
> $$\phi'' - \xi^2 \phi - R\Pi' = 0 \qquad (3)$$
> $$i(\alpha f + \beta h) + \phi' = 0 \qquad (4)$$
>
> Here
>
> $$\xi^2 = \gamma^2 + iRv, \qquad \gamma^2 = \alpha^2 + \beta^2, \qquad v \equiv \alpha U + \beta W - \omega$$
>
> where v is not to be confused with v of equation (12.2), and primes denote differentiation with respect to Y

We can obtain a single equation for ϕ as follows:
Multiply equation (1) by $i\alpha$, equation (2) by $i\beta$, and add

$$(i\alpha f + i\beta h)'' - (\gamma^2 + iRv)(i\alpha f + i\beta h) - iRv'\phi + R\gamma^2\Pi = 0. \qquad (6)$$

Substituting for $i\alpha f + i\beta h$ from equation (4),

$$\left[\frac{d^2}{dY^2} - (\gamma^2 + iRv)\right]\phi' + iR\frac{dv}{dY}\phi - R\gamma^2\Pi = 0.$$

Differentiating with respect to Y,

$$\left[\frac{d^2}{dY^2} - \xi\right]\phi'' - iR\frac{dv}{dY}\phi' + iR\frac{dv}{dY}\phi' + iR\frac{d^2v}{dY^2}\phi - R\gamma^2\Pi' = 0. \qquad (7)$$

From equation (3), multiplying both sides by γ^2, we get

$$\gamma^2\left[\frac{d^2}{dY^2} - (\gamma^2 + iRv)\right]\phi - R\gamma^2\Pi' = 0.$$

Subtracting from equation (7), we have

$$\left[\frac{d^2}{dY^2} - \xi\right](\phi'' - \gamma^2\phi) + iR\frac{d^2v}{dY^2}\phi = 0$$

or

$$\left[\frac{d^2}{dY^2} - \gamma^2\right]^2 \phi - iR[v(\phi'' - \gamma^2\phi) - v''\phi] = 0. \qquad (8)$$

12.5 Orr–Sommerfeld equation for two- and three-dimensional flows

For three-dimensional, incompressible flows, the Orr–Sommerfeld equation is

$$\phi^{iv} - 2\gamma^2 \phi'' + \gamma^4 \phi - iR[v(\phi'' - \gamma^2 \phi) - v''\phi] = 0 \qquad (1)$$

For two-dimensional flows, the Orr–Sommerfeld equation is

$$\phi^{iv} - 2\alpha^2 \phi'' + \alpha^4 \phi - iR[(\alpha U - \omega)(\phi'' - \alpha^2 \phi) - \alpha U''\phi] = 0 \qquad (2)$$

The boundary conditions are, for unbounded shear flows

$$Y = 0; \quad \phi = \phi' = 0 \quad \text{(no-slip)}$$
$$Y = Y_e; \quad \phi'' - \gamma^2 \phi + (\gamma + \xi)(\phi' + \gamma\phi) = 0 \qquad (3)$$
$$\phi''' - \gamma^2 \phi' + \xi(\phi'' - \gamma^2 \phi) = 0$$

with suitable choice of Y_e

For bounded shear flows the second condition is replaced by another no-slip condition

Equation (8) in the previous panel is called the Orr–Sommerfeld equation for three-dimensional flows and can be written in the form given by equation (1) in the above panel.

Note that the quantities U, W, ω, α, β and R are all dimensionless. Primes denote differentiation with respect to Y, which is also dimensionless ($\equiv y/l$).

For two-dimensional flows, where

$$\gamma^2 = \alpha^2 \quad \text{and} \quad v = \alpha U - \omega,$$

we can write the Orr–Sommerfeld equation in the form shown by equation (2) in the above panel.

Sometimes, the dimensionless radian frequency ω is replaced by

$$\omega = c\alpha, \qquad \bar{c} = c/U_0.$$

The quantity $c (\equiv c_r + ic_i)$ is the complex phase velocity of the disturbance.

The Orr–Sommerfeld equation is subject to the boundary conditions given in equation (3). For further details, see Keller (1974). In equation (3), values of γ and ξ must be chosen such that their real parts are positive.

12.6 Procedure for computing the eigenvalues of the Orr–Sommerfeld equation for spatial amplification

Two-dimensional stability
a. Velocity profile U is a function of y for given x by local similarity

b. $\beta = 0$ so that,
$$\phi_0'(\alpha, R, \omega) = 0$$
In spatial amplification theory, ϕ_0' depends on four scalars

Three-dimensional stability
a. Velocity profiles U, W are functions of y for a given x, z by local similarity

b. $\beta \neq 0$ and $\phi_0'(\alpha, \beta, R, \omega) = 0$ but α and β are related by the condition that $d\alpha/d\beta$ is real. As in two dimensions, ϕ_0' depends on four scalars with this condition

In stability analyses, it is useful to define two separate concepts; *spatial amplification* in which α and β are complex and ω is real, and *temporal amplification* in which ω is complex and α and β are real. In the spatial amplification theory, the amplitude varies with x and z as $(\exp(-\alpha_i X - \beta_i Z))$ and in the temporal amplification theory, it varies with time as $\exp(\omega_i T)$. In this chapter we are concerned with *spatial amplification*, except when explicitly stated.

Since the Orr–Sommerfeld equation and its boundary conditions are homogeneous, the trivial solution $\phi(y) = 0$ is valid for all values of R, α, β and ω. To compute the eigenvalues and eigenfunctions, we drop the boundary condition $\phi'(0) = 0$, replace it by the condition $\phi''(0) = 1$ and vary the appropriate combinations of α, β, R, ω until $\phi'(0)$ ($\equiv \phi_0'$) $= 0$; that is,

$$\phi_0'(\alpha, \beta, R, \omega) = 0.$$

Note that ϕ_0' depends on six scalars, α_r, α_i, β_r, β_i, R and ω which are real. We can only compute for two scalars and additional relations between the scalars may be needed.

The procedure for computing the eigenvalues of the Orr–Sommerfeld equation for two-dimensional flows is somewhat easier than the procedure for three-dimensional flows. In the latter case, the added condition, namely $d\alpha/d\beta$ being real, makes the eigenvalue hunting scheme difficult.

Sometimes it is assumed that the relation between α and β can be obtained by assuming θ, the angle which the disturbance makes with the x-axis as it propagates, i.e.

$$\tan \theta = \beta/\alpha. \tag{1}$$

For disturbances initiated at some point in the flow field and evolving downstream, equation (1) does not give the correct relation between α and β for the following reason. We can think of the disturbances as groups of waves with a particular frequency ω_0 radiating in all directions. In order to select those which make a significant contribution along a line inclined at an angle θ to the mainstream, we must remember that, generally, adjacent members of the group cancel each other because of the rapid oscillations. The only waves which make a significant contribution to the disturbance in the direction θ have wave numbers near to that with the property that as α and β vary, the quantity $\alpha X + \beta Z - \omega_0 T$ is stationary when $Z = X \tan \theta$, i.e.

$$X + \frac{d\beta}{d\alpha} Z = 0.$$

As a result the disturbance angle θ is computed from

$$\tan \theta = -\frac{d\alpha}{d\beta}.$$

For further details, see Cebeci and Stewartson (1980a, b) and Nayfeh (1980).

12.7 Numerical solution of the Orr–Sommerfeld equation for two-dimensional flows

> Reduce the system to first order. Let
> $$f = \phi'$$
> $$s = f' - \gamma^2 \phi$$
> $$g = s' \qquad (1)$$
> so that the Orr–Sommerfeld equation can be written as
> $$g' - \xi^2 s + iRv''\phi = 0$$
> **Boundary conditions**
> Wall: $\quad \phi = 0, \quad s = 1$
>
> Edge: $\quad g + \xi s = 0, \; f + \dfrac{s}{\gamma + \xi} + \gamma \phi = 0 \qquad (2)$
>
> Note f, ϕ, g, s are all complex and $\beta = 0$

Over the years, several numerical methods have been developed to solve the Orr–Sommerfeld equation for two-dimensional flows. Here we describe an efficient and accurate numerical method developed by Cebeci and Keller (1977) for solving this ordinary differential equation that uses the Box method described earlier.

We first reduce the Orr–Sommerfeld equation given by equation (1) in panel 12.5, to a system of first-order equations and write its usual and altered boundary conditions in terms of new variables as shown in the above panel. Step 2 of panel 5.11 or 6.3 is straightforward and Step 3 is not needed since the equation is linear.

The difference equations and altered boundary conditions, in matrix-vector form, can be written with the A_j, B_j, C_j matrices having 4×4 blocks.

$$\mathbf{A} \Delta = \mathbf{r}. \qquad (3)$$

In general, the solution of this equation for two-dimensional flows depends on five scalars, α, R and ω; that is

$$\Delta = \Delta(\alpha, R, \omega). \qquad (4)$$

Thus, we fix three scalars and seek the remaining two scalars to satisfy the missing boundary condition $f(0) \equiv f_0 = 0$, that is

$$f_0(\alpha, R, \omega) = 0. \qquad (5)$$

In spatial amplification theory the solution depends on four scalars, α_i, α_r, R, and ω(real). A similar procedure may be used if β is nonzero but prescribed, and may also be used with the roles of α and β interchanged.

12.8 Computation of eigenvalues and eigenfunctions on the neutral curve for two-dimensional flows

- Fix R, set $\alpha_i = 0$; guess α_r, ω. For a given U, U'', express equations (1), (2) of panel 12.2 in the form given by equation (3) and solve to get f_0. The next set of values of α_r and ω can be obtained by applying Newton's method to equation (5). With $f_0 \equiv f = f_r + if_i$

$$(f_r)^\nu + \left(\frac{\partial f_r}{\partial \alpha_r}\right)^\nu \delta\alpha_r^\nu + \left(\frac{\partial f_r}{\partial \omega}\right)^\nu \delta\omega^\nu = 0$$

$$(f_i)^\nu + \left(\frac{\partial f_i}{\partial \alpha_r}\right)^\nu \delta\alpha_r^\nu + \left(\frac{\partial f_i}{\partial \omega}\right)^\nu \delta\omega^\nu = 0$$

- Solve for

$$\delta\alpha_r^\nu, \quad \delta\omega^\nu$$

- Repeat with the new values of α_r, ω until convergence is achieved

To illustrate the calculation of the eigenvalues for a two-dimensional flow, let us assume that we want to compute a neutral stability curve for a given flow. We can fix R, and since $\alpha_i = 0$, we only need to solve for α_r and ω. For assumed values of α_r, ω we solve equations (3) of panel 12.7 to find f_0. To get the next values of $\alpha_r^{\nu+1}$ and $\omega^{\nu+1}$, we use Newton's method and write equation (5) of the previous page in the form shown above. Solving for $\delta\alpha_r^\nu$ and $\delta\omega^\nu$, we get

$$\delta\alpha_r^\nu = \left[(f_i)^\nu \left(\frac{\partial f_r}{\partial \omega}\right)^\nu - (f_r)^\nu \left(\frac{\partial f_i}{\partial \omega}\right)^\nu\right]/D_0,$$

$$\delta\omega^\nu = \left[(f_r)^\nu \left(\frac{\partial f_i}{\partial \alpha_r}\right)^\nu - (f_i)^\nu \left(\frac{\partial f_r}{\partial \alpha_r}\right)^\nu\right]/D_0,$$

where

$$D_0 = \left(\frac{\partial f_r}{\partial \alpha_r}\right)^\nu \left(\frac{\partial f_i}{\partial \omega}\right)^\nu - \left(\frac{\partial f_i}{\partial \alpha_r}\right)^\nu \left(\frac{\partial f_r}{\partial \omega}\right)^\nu.$$

Note

$$\alpha_r^{\nu+1} = \alpha_r^\nu + \delta\alpha_r^\nu,$$
$$\omega^{\nu+1} = \omega^\nu + \delta\omega^\nu.$$

To evaluate the derivatives of f_r and f_i, we need only differentiate Δ in equation (3) of panel 12.7. Since **r** is independent of α_r and ω, we get

$$\mathbf{A}\left(\frac{\partial \Delta}{\partial \alpha_r}\right)^v = -\left(\frac{\partial \mathbf{A}}{\partial \alpha_r}\right)^v (\Delta)^v,$$

$$\mathbf{A}\left(\frac{\partial \Delta}{\partial \omega}\right)^v = -\left(\frac{\partial \mathbf{A}}{\partial \omega}\right)^v (\Delta)^v.$$

Note that, to solve these variational equations, we need only compute new **r** since the coefficient matrix **A** has already been computed.

12.9 Computation of eigenvalues and eigenfunctions on the "zarf" for three-dimensional flows

- Fix ω and insist that $\alpha_i = \beta_i = 0$. Select a value for α_r and guess R, β. For given U, W profiles, solve the system of panel 12.7 to get f_0. To find R, β_r that satisfies $f_0 = 0$, use Newton's method, i.e.

$$(f_r)^v + \left(\frac{\partial f_r}{\partial \beta_r}\right)^v \delta \beta_r + \left(\frac{\partial f_r}{\partial R}\right)^v \delta R = 0$$

$$(f_i)^v + \left(\frac{\partial f_i}{\partial \beta_r}\right)^v \delta \beta_r + \left(\frac{\partial f_i}{\partial R}\right)^v \delta R = 0$$

- Compute $\partial f/\partial \beta_r$, $\partial f/\partial R$ from the variational equations. Then compute $d\alpha_r/d\beta_r$, $d\alpha_i/d\beta_r$. If $d\alpha_i/d\beta_r \neq 0$, then increment α_r and repeat the above procedure

To illustrate the calculation of the eigenvalues for a three-dimensional flow, let us assume that we want to compute the zarf[1] (see Cebeci and Stewartson, 1980a) on which also $\beta_i = 0$, $\alpha_i = 0$. The solution of equation (4) of panel 12.7 now depends on the four scalars, α_r, β_r, ω and R. If we fix ω, then for given U, W profiles and for a specified value of α_r, we can compute values of R, β_r (as shown in the above panel) which satisfy $f_0 = 0$.

We compute $\partial f/\partial \beta_r$, $\partial f/\partial R$ from variational equations obtained by differentiating the difference equations with respect to β_r, R, respectively.

To see whether $d\alpha/d\beta$ is real, we use

$$\frac{\partial f_r}{\partial \beta_r} + \frac{\partial f_r}{\partial \alpha_r}\frac{\partial \alpha_r}{\partial \beta_r} + \frac{\partial f_r}{\partial \alpha_i}\frac{\partial \alpha_i}{\partial \beta_r} = 0, \qquad \frac{\partial f_i}{\partial \beta_r} + \frac{\partial f_i}{\partial \alpha_r}\frac{\partial \alpha_r}{\partial \beta_r} + \frac{\partial f_i}{\partial \alpha_i}\frac{\partial \alpha_i}{\partial \beta_r} = 0. \qquad (2)$$

Since $\partial f/\partial \beta_r$, $\partial f/\partial \alpha_r$, $\partial f/\partial \alpha_i$ are all known (they are computed by solving variational equations obtained by differentiating the difference equations with respect to β_r, α_r and α_i), we can compute $\partial \alpha_r/\partial \beta_r$, $\partial \alpha_i/\partial \beta_r$ from equation (2). If $d\alpha/d\beta$ is not real, then we increment α_r and repeat the above procedure until we find a set of values of α_r, β_r, R for a given ω such that $\partial \alpha_i/\partial \beta_r = 0$.

---

[1] lit. envelope (Turkish). This is the locus of the points of minimum R on the neutral curves for fixed ω in spatial amplification theory and the envelope of neutral curves in temporal stability theory.

12.10 Results for Blasius flow (temporal stability)

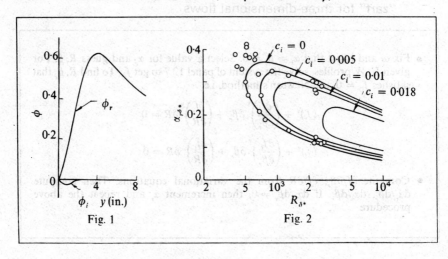

Fig. 1 Fig. 2

As a sample test case, we compute the Blasius flow. In this case the dimensionless velocity profile U in the Orr–Sommerfeld equation is simply f' (Blasius function) with the edge velocity U_e chosen to be U_0. A convenient length scale is δ^*, so that $Y = y/\delta^*$. The second derivative U'' is calculated from

$$U'' = f'''\,(\eta_e - f_e)^2.$$

Figure 1 of the above panel shows the real and imaginary parts of the eigenfunction, ϕ, corresponding to the lower branch of the neutral curve at $R_{\delta^*} = 902$ and Fig. 2 shows a stability diagram for Blasius flow obtained by using the temporal amplification theory for values of $c_i = 0$, 0·05, 0·01 and 0·018. The figure also shows the experimental data of Schubauer and Skramstad (1947). Similar results were also obtained by Kaplan (1964), Wazzan et al. (1968), Radbill and McCue (1970), and others.

For stability diagrams for other Falkner–Skan velocity profiles, the reader is referred to a comprehensive report by Wazzan et al. (1968).

12.11 Results for a rotating disk

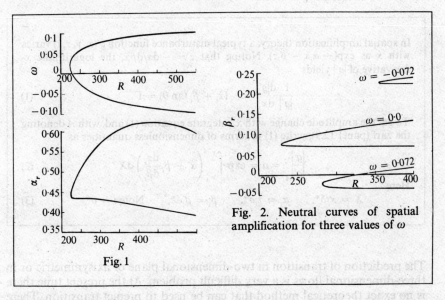

Fig. 1

Fig. 2. Neutral curves of spatial amplification for three values of ω

As an example for a 3-D flow, we now show the results for a rotating disk, see Cebeci and Stewartson (1980a). The velocity profiles U and V were obtained by solving the well-known similarity equations for a rotating disk.

Figure 1 shows the zarf for ω and α_r as a function of $R(\equiv a\sqrt{\omega/\nu})$. It was found that the critical Reynolds number $R_c = 176$ was close to Brown's results (1961); associated parameters are $\alpha_c = 0.348$, $\beta_c = 0.031$, $\omega_c = 0.017$, $\partial\alpha/\partial\beta = 6.95$.

Figure 2 shows the neutral curve of (β_r, R) for various values of ω. The value of $\partial\alpha/\partial\beta$ (not shown here but given in Cebeci and Stewartson, 1980a) is always >4 indicating that these disturbances travel almost parallel to the negative tangential direction. For additional discussions and comparisons, see Cebeci and Stewartson (1980a).

12.12 Transition prediction by the e^9-method, 1

> In spatial amplification theory, a typical disturbance function $g'(x, y, z, t)$ varies with x as $\exp(-\hat{\alpha}_i x - \hat{\beta}_i z)$. Noting that $z = -d\alpha/d\beta x$, the logarithmic x-derivative of $|g'|$ yields
>
> $$\frac{1}{|g'|}\frac{d|g'|}{dx} = -(\hat{\alpha}_i + \hat{\beta}_i \tan \theta) = \Gamma \qquad (1)$$
>
> To get the amplitude change with x, integrate equation (1) and, with A denoting the zarf (panel 12.9), write (1) in terms of dimensionless quantities as
>
> $$\left|\frac{g'}{g'}\right|_A = a = \exp \int_{X_A}^{X} \left(\alpha_i - \beta_i \frac{d\alpha_r}{d\beta_r}\right) dX \qquad (2)$$
>
> Here
> $$X = x/\delta^*, \qquad \alpha_i = \hat{\alpha}_i \delta^*, \qquad \beta_i = \hat{\beta}_i \delta^*, \qquad \text{Note } l = \delta^* \qquad (3)$$

The prediction of transition in two-dimensional plane or axisymmetric or in three-dimensional flows is a very difficult problem. At the present time there is no exact theoretical method that can be used to predict transition. There is one empirical method that seems to work reasonably well for two-dimensional and axisymmetric flows. This is the so-called e^9-method, which utilizes the stability theory and some experimental results. It was first used by Smith and Gamberoni (1956) and by Van Ingen (1956). The basic assumption is that transition starts when a small disturbance introduced at the critical Reynolds number has amplified by a factor of e^9 or about 8000. Strictly speaking, the criterion is that amplification predicted by the linear theory should have reached e^9. This method is, of course, wholly empirical. It is probably better to regard the method as an empirical correlation for the length of the transition region of a boundary layer in a low-turbulence stream since a high amplification rate will usually imply a short transition region. The method requires a knowledge of the disturbance amplitude, $g(y)$, at each x-station in a boundary layer. Either temporal or spatial amplification theory can be used, but the latter makes far more physical sense since it is concerned with the evolution of a disturbance of fixed frequency as it moves downstream from a source. Hence it may be generalized to permit R to vary and still gives a good estimate of the growth in the amplitude of the disturbance (Gaster, 1975). The temporal amplification is strictly concerned with the evolution of an initial disturbance in its neighbourhood.

The e^9-method requires the change of the amplitude of the disturbance with x. To compute that for a two-dimensional flow, we use equation (2) and the following procedure.

1. For a given external velocity distribution and freestream Reynolds number, solve the thin shear-layer equations to get U, U'' and $R(\equiv u_e \delta^*/v)$.

Stability and Transition

With $R =$ given, and with spatial amplification theory, solve the Orr–Sommerfeld equation to compute ω, $\alpha_r (\equiv \hat{\alpha}_r \delta^*)$ on the neutral stability curve. Determine physical frequency $\hat{\omega} = \omega u_e/\delta^*$.

2. Go to next station, solve the thin shear-layer equations to obtain a new R and, with ω known from

$$\omega = (\hat{\omega}/u_e)\delta^*,$$

solve Orr–Sommerfeld equation to get α_r, α_i.

3. Go to next station and repeat 2.

The stability calculations usually start at the lower branch of the neutral stability curve by computing ω and α_r. From ω we compute the physical dimensional $\hat{\omega}$ which is kept constant in the subsequent x-stations.

For three-dimensional flows the procedure is similar to that used for two-dimensional flows, although the calculation of eigenvalues subject to $d\alpha/d\beta$ being real complicates the hunting scheme and makes the calculation of the amplification rates, $\alpha_i - \beta_i \, d\alpha_r/d\beta_r$ difficult. The procedure is

1. For a given external velocity distribution and freestream Reynolds number, solve the thin shear-layer equations to get U, W, U'', W'' and R. With R given, use spatial amplification theory to compute ω on the zarf.

2. Go to next station, solve the thin shear-layer equations to get a new R and with ω known from

$$\omega = \hat{\omega}_A/u_e \delta^*$$

solve Orr–Sommerfeld equation to get α_i, α_r, β_r, β_i. In this calculation strictly $d\alpha/d\beta$ should be varied to maximise $-(\alpha_i + \beta_i \tan \theta)$ but an alternative procedure is to keep $d\alpha/d\beta$ constant at the value it takes on the zarf. The loss of accuracy is believed to be small, at least for the rotating disc problem. However, this needs to be explored further.

12.13 Transition prediction by the e^9-method, 2

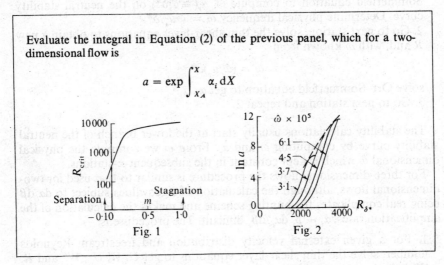

Evaluate the integral in Equation (2) of the previous panel, which for a two-dimensional flow is

$$a = \exp \int_{X_A}^{X} \alpha_i \, dX$$

Fig. 1

Fig. 2

The stability calculations begin at a $R_{\delta^*} > R_{\delta^* \text{crit}}$ on the lower branch of the neutral stability curve. Figure 1 of the above panel shows the variation of R_{crit} with pressure gradient parameter $m (\equiv X/U_e \, dU_e/dX)$. Let us consider the Blasius flow. In this case a choice of $R = 600$ is quite satisfactory to initiate the stability calculations, since $R_{\text{crit}} = 520$. For a given value of ω determined at one R at the neutral stability curve, we solve the Orr–Sommerfeld equation for several values of R and compute the variation of $\ln a$ with X. The calculations stop when the variation in $\ln a$ with X becomes negligible. After that, a new set of calculations is made for a different value of frequency, ω. Figure 2 of the above panel shows the computed amplification factors for ω ranging from 0.61×10^4 to 0.310×10^4 rad s^{-1}. The envelope of these curves represents the maximum amplification factor, $A(R)$. Transition is assumed to start when $\ln A = 9$ or 10. According to the calculations of Cebeci and Keller (1977), at $(R_x)_{\text{tr}} = 2.84 \times 10^6$, $\ln A = 8.8$. Similar results were also obtained by Smith and Gamberoni (1956) and by Jaffe et al. (1970) with slightly different values of $\ln A$.

Stability and Transition

12.14 Results for a body of revolution: prolate spheroid at zero incidence

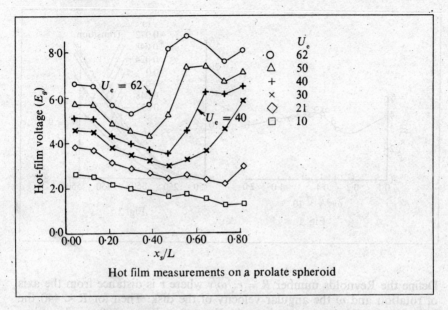

Hot film measurements on a prolate spheroid

Although the e^9-method has been used for some time for predicting the transition on two-dimensional flows and axisymmetric bodies, it has not been made quite clear whether this method predicts the onset of transition or the end of transition. Since the transitional region is a function of Reynolds number and becomes more and more pronounced at lower Reynolds numbers, it is important to establish which location of the transition region this empirical method predicts. According to our recent calculations for a body of revolution at zero incidence, it appears that the e^9-method predicts the *onset* of transition. Experimental results for six Reynolds numbers are shown in the above figure. The data, hot-film voltages representing surface shear stress, are due to Meier and Kreplin (1980) and the body of revolution is a prolate spheroid 240 cm long with a thickness ratio of 1/6. For $U_e = 62 \text{ m s}^{-1}$, which corresponds to a length Reynolds of $9 \cdot 6 \times 10^6$, the predicted transition location is 0·405, agreeing well with the experimental value of 0·40. For $U_e = 40 \text{ m s}^{-1}$, the computed transition location is 0·595, and the experimental (the point at which the hot-film voltage starts to rise) location is somewhere around 0·50 to 0·55.

12.15 Results for a rotating disk

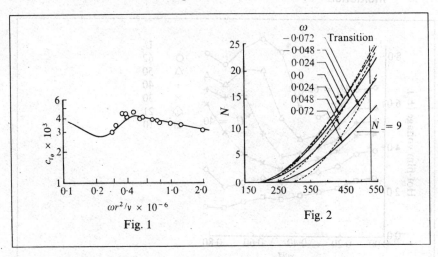

Fig. 1

Fig. 2

Define the Reynolds number $R = r\sqrt{\omega/\nu}$ where r is distance from the axis of rotation and ω the angular velocity of the disk. Then for $R < 440$ the boundary layer is laminar and steady. For $440 < R < \sim 530$ spiral vortices appear in the flow which become unsteady although the moment coefficient is still given by laminar theory. At about $R = 530$ there is a rapid rise in drag and the flow becomes fully turbulent.

Figure 1 in the above panel shows a comparison between the experimental data of Cham and Head (1969) and the numerical calculations of Cebeci and Abbott (1975), indicated by the solid line. The calculations were made by solving the governing equations for a rotating disk by the Box method and using the Cebeci and Smith model for the Reynolds stress terms. The laminar flow calculations were made up to Reynolds number 530; the turbulent flow calculations using the eddy viscosity model modified for the transitional region, were then made up to the rest of the flow. As is seen, the agreement between calculations and experimental data is very good.

If the empirical e^9 method predicted the onset of transition for all flows over smooth bodies with low ambient turbulence then a more sensible approach is to regard it as a special case of the e^n method, computing the value of n for each flow under investigation. For the rotating disk, we use equation (2) of panel 12.12, starting at the zarf and holding $d\alpha/d\beta$ constant at the value it takes there. In Fig. 2 we display the variation of $\ln a$ with R for various ω and obtain a value $\simeq 23$ for n. The variation is large, illustrating the limitation of the method and indicating the need for further development, particularly in regard to three-dimensional aspects.

13 Wings

13.1 Purposes and outline of chapter

- To describe the formulation of a general method for computing the properties of the three-dimensional boundary layers on finite wings
- To describe the numerical procedure used to solve the governing equations with and without flow reversal in the spanwise velocity profile
- To discuss the inviscid–viscous interaction for three-dimensional flows by using the displacement surface and blowing approaches

Boundary-layer calculations were described in Chapters 6 to 9 and, particularly in Chapter 8, consideration was given to three-dimensional boundary layers. In this chapter, we return to this topic in the particular context of wings.

The emphasis of this chapter is on the solution of the three-dimensional boundary-layer equations appropriate to arbitrary-shaped wings. This requires, as indicated on the following panel, various components which are described on the succeeding pages with particular emphasis given to the numerical method. In this discussion, it may be presumed that the external velocity distribution is prescribed and known. In practice, however, and as introduced in previous chapters, the potential-flow equations must be solved and the result iterated with the boundary-layer solutions to ensure that the inviscid–viscous interaction is correctly represented. Comments on this interaction are provided on panels 13.11 and 13.12.

Two aspects of wing design which are important and are not considered here are the representation of supercritical wings, with their suction-side shock wave and related numerical and turbulence-model problems, and the inviscid–viscous interaction in the wake. Both are topics for further research which is in progress and shows that the numerical problems are greater than those associated with modelling but can be overcome. The paper by Lynch (1978) reports progress on these two problems, respectively.

13.2 Essential components of a 3-D boundary-layer method

The development of a three-dimensional boundary-layer method requires
1. The specification of the external velocity distribution
   (a) either experimentally or (b) theoretically
2. An appropriate coordinate system and the calculation of its geometric parameters

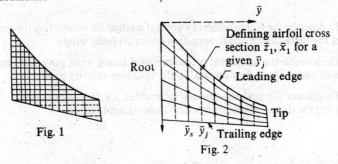

Fig. 1

Fig. 2

3. The solution of the boundary-layer equations with a suitable turbulence model and an accurate numerical method

As an example of the calculation of three-dimensional steady flows, we now consider three-dimensional boundary layers on arbitrary wings. The development of a corresponding method requires the three items discussed in the above panel.

For low speeds, the external velocity distribution can be obtained from the Hess potential flow method (1979): for transonic flows, it can be obtained from the Jameson–Caughey method (see Caughey and Jameson 1979). When the pressure distribution is known from experiments, the external velocity distribution can be obtained by an approximate procedure such as that described by Cebeci, Kaups and Ramsey (1977).

The choice of an appropriate coordinate system is very important, especially with three-dimensional flows on wings. To illustrate this point, consider an orthogonal system in which the orthogonals are constructed at constant percent chord stations. With this system, as shown in Fig. 1 of the panel, the orthogonals start from the wing root and intersect the leading edge, leaving large portions near the trailing edge uncovered. This is especially true for a wing with a sharp-tipped trailing edge; it is less so for a round-tipped trailing edge but there is still a large area of the wing where the orthogonals are sparse.

An ideal coordinate system is a nonorthogonal system in which the coordinate lines in the spanwise direction are constructed by joining the constant percent-chord lines. It was used by Cebeci, Kaups and Ramsey (1977) and is shown in Fig. 2 of the panel.

Wings

13.3 Boundary-layer equations for a three-dimensional non-orthogonal system

$$\frac{\partial}{\partial x}(\rho U h_2 \sin\theta) + \frac{\partial}{\partial z}(\rho W h_1 \sin\theta) + \frac{\partial}{\partial y}(\overline{\rho V} h_1 h_2 \sin\theta) = 0 \qquad (1)$$

$$\rho \frac{U}{h_1}\frac{\partial U}{\partial x} + \rho \frac{W}{h_2}\frac{\partial U}{\partial z} + \overline{\rho V}\frac{\partial U}{\partial y} - \rho \cot\theta\, K_1 U^2 + \rho \csc\theta K_2 W^2 + \rho K_{12} UW$$

$$= -\frac{\csc^2\theta}{h_1}\frac{\partial P}{\partial x} + \frac{\cot\theta \csc\theta}{h_2}\frac{\partial P}{\partial z} + \frac{\partial}{\partial y}\left(\mu\frac{\partial U}{\partial y} - \rho\overline{uv}\right) \qquad (2)$$

$$\rho \frac{U}{h_1}\frac{\partial W}{\partial x} + \rho \frac{W}{h_2}\frac{\partial W}{\partial z} + \overline{\rho V}\frac{\partial W}{\partial y} - \rho \cot\theta K_2 W^2 + \rho \csc\theta K_1 U^2 + \rho K_{21} UW$$

$$= \frac{\cot\theta \csc\theta}{h_1}\frac{\partial}{\partial x} - \frac{\csc^2\theta}{h_2}\frac{\partial}{\partial z} + \frac{\partial}{\partial y}\left(\mu\frac{\partial W}{\partial y} - \rho\overline{wv}\right) \qquad (3)$$

The governing boundary-layer equations for a three-dimensional non-orthogonal system are shown in the panel. Here $\overline{\rho V} = \rho V + \overline{\rho v}$, and h_1 and h_2 are metric coefficients which are functions of x and z. $\theta(x, z)$ represents the angle between the coordinate lines x and z. For an orthogonal system $\theta = \pi/2$. The parameters K_1 and K_2 are known as the geodesic curvatures of the curves $z = \text{const.}$ and $x = \text{const.}$, respectively. They are given by

$$K_1 = \frac{1}{h_1 h_2 \sin\theta}\left[\frac{\partial}{\partial x}(h_2 \cos\theta) - \frac{\partial h_1}{\partial z}\right], \quad K_2 = \frac{1}{h_1 h_2 \sin\theta}\left[\frac{\partial}{\partial z}(h_1 \cos\theta) - \frac{\partial h_2}{\partial x}\right].$$

The parameters K_{12} and K_{21} are defined by

$$K_{12} = \frac{1}{\sin\theta}\left[-\left(K_1 + \frac{1}{h_1}\frac{\partial\theta}{\partial x}\right) + \cos\theta\left(K_2 + \frac{1}{h_2}\frac{\partial\theta}{\partial z}\right)\right],$$

$$K_{21} = \frac{1}{\sin\theta}\left[-\left(K_2 + \frac{1}{h_2}\frac{\partial\theta}{\partial z}\right) + \cos\theta\left(K_1 + \frac{1}{h_1}\frac{\partial\theta}{\partial x}\right)\right].$$

Before the equations can be solved, it is necessary to compute the metric coefficients and geodesic curvatures, i.e. the geometric parameters of the coordinate system. A general computer program that performs this task is described by Cebeci, Kaups and Ramsey (1977).

13.4 An algebraic eddy-viscosity formulation for 3-D flows

$$v_t = \begin{cases} = L^2\left[\left(\frac{\partial U}{\partial y}\right)^2 + \left(\frac{\partial W}{\partial y}\right)^2 + 2\cos\theta\left(\frac{\partial U}{\partial y}\right)\left(\frac{\partial W}{\partial y}\right)\right]^{1/2} \\ = 0{\cdot}0168\left|\int_0^\infty (U_{te} - U_t)\,dy\right| \end{cases} \quad (1)$$

$$L = 0{\cdot}4y[1 - \exp(-y/A)] \quad (2)$$

$$U_t = (U^2 + W^2 + 2\cos\theta UW)^{1/2} \quad (3)$$

In addition to the calculation of the geometric parameters of the coordinate system it is necessary to have a model for the Reynolds shear stress terms, namely, for $-\rho\overline{uv}$, $-\rho\overline{vw}$. A generalization of the algebraic eddy viscosity formulation used by Cebeci and Smith (1974) for two-dimensional flows is given above for three-dimensional flows. This formulation has been appraised by comparison with the available experimental data and, with the exception of data of van den Berg et al., (1975), has been found to give satisfactory results; see Cebeci (1974, 1975), Cebeci, Kaups and Moser (1976), Cebeci et al. (1973) and Cebeci, Kaups and Ramsey (1977) for further details. See also panel 3.6.

It might be argued that three-dimensional flows of this type require the use of a turbulence model which allows for the transport of turbulence quantities. The paper by Cebeci and Meier (1979), with the examination of the models of Cebeci and Smith (1974), Bradshaw et al. (1967) and Jones and Launder (1972), suggests that this is incorrect and that the simplest model should be used. The lack of experimental information for wing-type flows and problems associated with transition in wind-tunnel experiments are two of the main reasons for this conclusion.

Wings

13.5 Initial conditions, 1

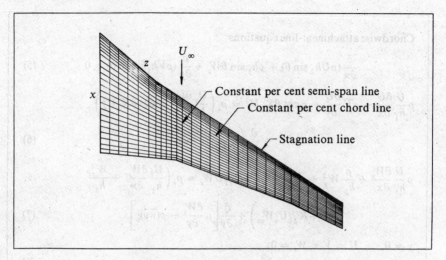

The boundary-layer calculations on the wing require initial conditions on the stagnation line and on the root section.

The initial conditions on the stagnation line can be obtained from those equations given in panel 13.3 by noting that on this line $U = 0$ and $\partial P/\partial x = 0$. This makes the x-momentum equation singular on that line. However, differentiation with respect to x yields a nonsingular equation. After performing the necessary differentiation for the x-momentum equation and taking advantage of appropriate symmetry conditions ($\partial W/\partial x = \partial V/\partial x = \partial^2 U/\partial x^2 = 0$), the governing *stagnation-line* equations may be written as

$$\rho h_2 \sin\theta\, U_x + \frac{\partial}{\partial z}(\rho h_1 \sin\theta\, W) + \frac{\partial}{\partial y}(\overline{\rho V} h_1 h_2 \sin\theta) = 0, \quad (1)$$

$$\rho \frac{U_x^2}{h_1} + \rho \frac{W}{h_2}\frac{\partial U_x}{\partial z} + \overline{\rho V}\frac{\partial U_x}{\partial y} + \rho K_{12} W U_x$$
$$= \rho_e\left(\frac{U_{xe}^2}{h_1} + \frac{W_e}{h_2}\frac{\partial U_{xe}}{\partial z} + K_{12} W_e U_{xe}\right) + \frac{\partial}{\partial y}\left[\mu \frac{\partial U_x}{\partial y} - \rho\overline{(uv)}_x\right], \quad (2)$$

$$\rho \frac{W}{h_2}\frac{\partial W}{\partial z} + \overline{\rho V}\frac{\partial W}{\partial y} - \rho \cot\theta\, K_2 W^2 = \rho_e\left(\frac{W_e}{h_2}\frac{\partial W_e}{\partial z} - \cot\theta\, K_2 W_e^2\right)$$
$$+ \frac{\partial}{\partial y}\left[\mu \frac{\partial W}{\partial y} - \rho\overline{(wv)}\right] \quad (3)$$

Here $U_x = \partial U/\partial x$, $U_{xe} = \partial U_e/\partial x$. The boundary conditions are

$$\begin{aligned} y = 0, & \quad U_x = V = W = 0, \\ y = \delta, & \quad U_x = U_{xe}(x,z), \quad W = W_e(x,z). \end{aligned} \quad (4)$$

13.6 Initial conditions, 2

Chordwise attachment-line equations

$$\frac{\partial}{\partial x}(\rho U h_2 \sin\theta) + \rho h_1 \sin\theta W_z + \frac{\partial}{\partial y}(\rho \overline{V} h_1 h_2 \sin\theta) = 0 \tag{5}$$

$$\rho \frac{U}{h_1}\frac{\partial U}{\partial x} + \rho\overline{V}\frac{\partial U}{\partial y} - \rho \cot\theta K_1 U^2 = \rho_e\left(\frac{U_e}{h_1}\frac{\partial U_e}{\partial x} - \cot\theta K_1 U_e^2\right)$$

$$+ \frac{\partial}{\partial y}\left(\mu\frac{\partial U}{\partial y} - \rho\overline{uv}\right) \tag{6}$$

$$\rho\frac{U}{h_1}\frac{\partial W_z}{\partial x} + \frac{\rho}{h_2}W_z^2 + \rho\overline{V}\frac{\partial W_z}{\partial y} + \rho K_{21}UW_z = \rho_e\left(\frac{U_e}{h_1}\frac{\partial W_{ze}}{\partial x} + \frac{W_{ze}^2}{h_2}\right.$$

$$\left.+ K_{21}U_e W_{ze}\right) + \frac{\partial}{\partial y}\left[\mu\frac{\partial W_z}{\partial y} - \rho\overline{(wv)}_z\right] \tag{7}$$

$y = 0, \quad U = V = W_z = 0;$

$y = \delta, \quad U = U_e(x,z), \quad W_z = W_{ze} \tag{8}$

The initial conditions on the root section can be started by using the infinite swept-wing assumptions or the attachment-line equations. Both are approximations to the wing-fuselage intersection and a proper treatment requires the solution of the corner-flow equations. The corner-flow problem has been considered in Chapters 10 and 11, and it should be clear that this involves skew-induced secondary flows and requires additional experimental and calculation work before the initial condition at the wing-root/fuselage intersection can be obtained from the "proper" equations. In the meantime and for sometime to come, the above approximations will be necessary.

The attachment-line equations are obtained in a manner similar to the stagnation-line equations. On this line W and $\partial P/\partial z$ are zero making the z-momentum equation singular. However, differentiation with respect to z yields a nonsingular equation. After performing the necessary differentiation for the z-momentum and taking advantage of the appropriate symmetry conditions ($\partial U/\partial z = \partial V/\partial z = 0$), we can write the governing chordwise attachment-line equations and their boundary conditions as shown in the panel.

Wings

13.7 Transformed equations, general case

$$(bf'')' + f''e - m_2(f')^2 - m_5 f'g' - m_8(g')^2 + m_{11}c =$$
$$m_{10} f' \frac{\partial f'}{\partial x} + m_7 g' \frac{\partial f'}{\partial z} \quad (4)$$

$$(bg'')' + g''e - m_4 f'g' - m_3(g')^2 - m_9(f')^2 + m_{12}c =$$
$$m_{10} f' \frac{\partial g'}{\partial x} + m_7 g' \frac{\partial g'}{\partial z} \quad (5)$$

$$e' = m_1 f' + m_6 g' + m_{10} \frac{\partial f'}{\partial x} + m_7 \frac{\partial g'}{\partial z} \quad (6)$$

$$\eta = 0, \ f' = g' = 0, e = 0 \text{ (no mass transfer)} \quad (7)$$
$$\eta = \eta_e, f' = 1, g' = W_e/U_{ref}$$

Here

$$f' = \frac{U}{U_e}, \qquad g' = \frac{W}{U_{ref}},$$

$$e = -\frac{s_1}{\sqrt{\rho_e \mu_e U_e s_1}} \overline{\rho V} - f' \frac{s_1}{h_1} \frac{\partial \eta}{\partial x} - g' \frac{U_{ref}}{U_e} \frac{s_1}{h_1} \frac{\partial \eta}{\partial z}$$

The boundary-layer equations can be solved when they are expressed either in physical coordinates or in transformed coordinates, and both systems have their own advantages. In three-dimensional flows, where the computer storage and time become important, transformed coordinates become necessary, as well as convenient, because they allow large steps to be taken in the streamwise and spanwise directions. In addition, they remove the singularity the equations have in physical coordinates at $x = 0$ and $z = 0$, and provide a convenient procedure for generating the initial conditions for laminar flows.

A convenient, useful transformation for three-dimensional boundary layers has been described by Cebeci, Kaups and Ramsey (1977) and is given by

$$x = x, \qquad z = z, \qquad d\eta = \left(\frac{U_e}{\rho_e \mu_e s_1}\right)^{1/2} \rho \, dy, \qquad s_1 = \int_0^x h_1 \, dx, \quad (1)$$

with a two-component vector potential such that

$$\rho U h_2 \sin\theta = \frac{\partial \psi}{\partial y}, \qquad \rho W h_1 \sin\theta = \frac{\partial \phi}{\partial y}, \qquad \overline{\rho V} h_1 h_2 \sin\theta = -\left(\frac{\partial \psi}{\partial x} + \frac{\partial \phi}{\partial z}\right), \quad (2)$$

and dimensionless ψ and ϕ are defined by

$$\psi = (\rho_e \mu_e U_e s_1)^{1/2} h_2 \sin\theta f(x, z, \eta),$$
$$\phi = (\rho_e \mu_e U_e s_1)^{1/2} U_{\text{ref}}/U_e h_1 \sin\theta g(x, z, \eta). \tag{3}$$

With this transformation, we can write the general equations of panel 13.3 and their boundary conditions as shown above. Here the coefficients m_1 to m_{12} denote dimensionless pressure gradients and fluid properties; they are given by

$$m_1 = \frac{1}{2}\left(1 + \frac{s_1}{h_1 U_e}\frac{\partial U_e}{\partial x}\right) + \frac{s_1(\rho_e \mu_e)^{-1/2}}{h_1 h_2 \sin\theta}\frac{\partial}{\partial x}(h_2 \sin\theta\sqrt{\rho_e \mu_e}),$$

$$m_2 = \frac{s_1}{h_1 U_e}\frac{\partial U_e}{\partial x} - s_1 K_1 \cot\theta,$$

$$m_3 = -s_1 \cot\theta K_2 \frac{U_{\text{ref}}}{U_e},$$

$$m_4 = s_1 K_{21}, \qquad m_5 = \frac{s_1}{h_2}\frac{U_{\text{ref}}}{U_e^2}\frac{\partial U_e}{\partial z} + K_{12} s_1 \frac{U_{\text{ref}}}{U_e},$$

$$m_6 = \frac{s_1}{h_1}\left(\sqrt{\rho_e \mu_e U_e s_1}\, h_2 \sin\theta\right)^{-1}\frac{\partial}{\partial z}\left(\sqrt{\rho_e \mu_e U_e s_1}\, h_1 \sin\theta \frac{U_{\text{ref}}}{U_e}\right),$$

$$m_7 = \frac{s_1}{h_2}\frac{U_{\text{ref}}}{U_e}, \qquad m_8 = s_1 K_2 \csc\theta \left(\frac{U_{\text{ref}}}{U_e}\right)^2, \qquad m_9 = s_1 K_1 \csc\theta \frac{U_e}{U_{\text{ref}}},$$

$$m_{10} = \frac{s_1}{h_1},$$

$$m_{11} = s_1\left[\frac{1}{U_e h_1}\frac{\partial U_e}{\partial x} + \frac{W_e}{U_e^2 h_2}\frac{\partial U_e}{\partial z} - \cot\theta K_1 + \csc\theta K_2 \left(\frac{W_e}{U_e}\right)^2 + K_{12}\frac{W_e}{U_e}\right],$$

$$m_{12} = \frac{s_1}{U_e U_{\text{ref}}}\left(\frac{U_e}{h_1}\frac{\partial W_e}{\partial x} + \frac{W_e}{h_2}\frac{\partial W_e}{\partial z} - \cot\theta K_2 W_e^2 + \csc\theta K_1 U_e^2 + K_{21} W_e U_e\right).$$

Wings

13.8 Transformed equations, stagnation line

$$(bf''')' + f''e - (f')^2 - m_5 f'g' + m_{11}c = m_7 g' \frac{\partial f'}{\partial z} \quad (4)$$

$$(bg''')' + g''e - m_3(g')^2 + m_{12}c = m_7 g' \frac{\partial g'}{\partial z} \quad (5)$$

$$e' = f' + m_6 g' + m_7 \frac{\partial g'}{\partial z} \quad (6)$$

$$\eta = 0, \quad f' = g' = e = 0$$
$$\eta = \eta_e, \quad f' = 1, \quad g' = 1, \quad g' = W_e/U_{ref} \quad (7)$$

To transform the stagnation-line equations given in panel 13.5, we define the similarity variable η by

$$d\eta = \left(\frac{U_{xe}}{\rho_e \mu_e h_1}\right)^{1/2} \rho \, dy, \quad (1)$$

the two-component vector potential by

$$\rho U_x h_2 \sin\theta = \frac{\partial \psi}{\partial y}, \quad \rho W h_1 \sin\theta = \frac{\partial \phi}{\partial y}, \quad \overline{\rho V} h_1 h_2 \sin\theta = -\left(\psi + \frac{\partial \phi}{\partial z}\right), \quad (2)$$

and the dimensionless ψ and ϕ by

$$\psi = (\rho_e \mu_e U_{xe} h_1)^{1/2} h_2 \sin\theta f(z, \eta), \quad \phi = (\rho_e \mu_e U_{xe} h_1)^{1/2} h_1 \sin\theta \frac{U_{ref}}{U_{xe}} g(z, \eta). \quad (3)$$

With these new variables, the stagnation-line equations and their boundary conditions can be written as shown in the panel.

Here m_3, m_5, m_6, m_{11} and m_{12} are:

$$m_3 = -h_1 K_2 \cot\theta \frac{U_{ref}}{U_{xe}},$$

$$m_5 = h_1 \frac{U_{ref}}{U_{xe}} \left(K_{12} + \frac{1}{h_2 U_{xc}} \frac{\partial U_{xe}}{\partial z}\right),$$

$$m_6 = [(\rho_e \mu_e U_{xe} h_1)^{1/2} h_2 \sin\theta]^{-1} \frac{\partial}{\partial z}\left(\sqrt{\rho_e \mu_e U_{xe} h_1}\, h_1 \sin\theta \frac{U_{ref}}{U_{xe}}\right),$$

$$m_7 = \frac{h_1}{h_2} \frac{U_{ref}}{U_{xe}},$$

$$m_{11} = 1 + \frac{h_1}{h_2}\frac{W_e}{U_{xe}^2}\frac{\partial U_{xe}}{\partial z} + h_1 K_{12}\frac{W_e}{U_{xe}}, \qquad (8)$$

$$m_{12} = \frac{h_1}{h_2}\frac{W_e}{U_{xe}U_{ref}}\frac{\partial W_e}{\partial z} - \frac{h_1\cot\theta}{U_{xe}U_{ref}} K_2 W_e^2.$$

13.9 Transformed equations, chordwise attachment line

$$(bf'')' + f''e - m_2(f')^2 + m_{11}c = \frac{s_1}{h_1} f' \frac{\partial f'}{x_1} \tag{3}$$

$$(bg'')' + g''e - m_4 f'g' - m_3(g')^2 - m_9(f')^2 + m_{12}c = \frac{s_1}{h_1} f' \frac{\partial g'}{\partial x} \tag{4}$$

$$e' = m_1 f' + m_6 g' + \frac{s_1}{h_1} \frac{\partial f'}{\partial x} \tag{5}$$

$$\eta = 0, \quad f' = g' = e = 0$$

$$\eta = \eta_e, \quad f' = 1, \quad g' = \frac{W_{ze}}{U_{ref}} \tag{6}$$

To transform the governing chordwise attachment-line equations, we use the transformed coordinates given by equation (1) of panel 13.7 and define the two-component vector potential by

$$\rho U h_2 \sin\theta = \frac{\partial \psi}{\partial y}, \quad W_z \rho h_1 \sin\theta = \frac{\partial \phi}{\partial y}, \quad \rho V h_1 h_2 \sin\theta = -\left(\frac{\partial \psi}{\partial x} + \phi\right), \tag{1}$$

with ψ and ϕ still given by equation (3) of panel 13.7. With these variables, the governing equations and their boundary conditions can be written as shown in the above panel.

The definitions of the coefficients m_1 to m_{12} in these equations are the same as those in equation (8) of panel 13.7 except that now

$$m_3 = \frac{s_1}{h_2} \frac{U_{ref}}{U_e}, \quad m_6 = m_3, \quad m_9 = 0,$$

$$m_{11} = \frac{s_1}{h_1} \frac{1}{U_e} \frac{\partial U_e}{\partial x} - s_1 \cot\theta K_1, \quad m_{12} = \frac{s_1}{U_{ref}} \left(\frac{1}{h_1} \frac{\partial W_{ze}}{\partial x} + \frac{W_{ze}^2}{U_e h_2} + K_{21} W_{ze}\right).$$

13.10 Regular Box, 1

$$U' = G \tag{1a}$$

$$W' = S \tag{1b}$$

$$(bG)' + eG - m_2 U^2 - m_5 UW - m_8 W^2 + m_{11}c = m_{10} U \frac{\partial U}{\partial x} + m_7 W \frac{\partial U}{\partial z} \tag{1c}$$

$$(bS)' + eS - m_4 UW - m_3 W^2 - m_9 U^2 + m_{12}c = m_{10} U \frac{\partial W}{\partial x} + m_7 W \frac{\partial W}{\partial z} \tag{1d}$$

$$e' = m_1 U + m_6 W + m_{10} \frac{\partial U}{\partial x} + m_7 \frac{\partial W}{\partial z} \tag{1e}$$

$$\eta = 0, \quad U = W = e = 0 \tag{2a}$$

$$\eta = \eta_\infty, \quad U = 1, \quad W = \frac{W_e}{U_{\text{ref}}} \tag{2b}$$

Note that for convenience, U and W are the nondimensional form for the velocity components.

Once the initial conditions in the spanwise and streamwise directions are computed, then we solve the equations for the general case and, for a given streamwise station, march in the spanwise direction. Depending on the complexity of the flow, we either use the Regular Box or the Characteristic Box, which are discussed here and on the following two pages.

Corresponding to step 1 of panel 5.11, the three-dimensional equations of panel 13.7 are written as five first-order equations by introducing the new dependent variables $G(x, z, \eta)$ and $S(x, z, \eta)$. With $f'(\equiv U/U_e)$ written as U and $g'(\equiv W/U_{\text{ref}})$ written as W, the first-order equations and their boundary conditions have the form shown on the panel.

The finite difference form of equation (1) may then be written, following step 2 of panel 5.11, in relation to the diagram of panel 8.3 and with t replaced by z. Again, we approximate equations (1a) and (1b) with central difference quotients and averages about the points $(x_i, z_n, \eta_{j-\frac{1}{2}})$. The difference equations which approximate equations (1c), (1d), (1e) are written about the midpoint $(x_{i-\frac{1}{2}}, z_{n-\frac{1}{2}}, \eta_{j-\frac{1}{2}})$. Further details are provided on the following page.

13.11   Regular Box, 2

Average of a quantity q at $(x_{i-\frac{1}{2}}, z_{n-\frac{1}{2}}, \eta_{j-\frac{1}{2}})$ is defined by

$$\bar{q}_{j-\frac{1}{2}} = \tfrac{1}{2}(\bar{q}_j + \bar{q}_{j-1})$$

where

$$\bar{q}_j = \tfrac{1}{4}(q_j^{i,n} + q^{i-1,n} + q_j^{i,n-1} + q^{i-1,n-1})$$

All parameters except e_j are averaged in this way, and $q^{i,n}$ is treated as the unknown. e_j is averaged as

$$\bar{e}_{j-\frac{1}{2}} = \tfrac{1}{2}(\bar{e}_j + \bar{e}_{j-1})$$

with $\bar{e}_j$ being treated as the *unknown*

In centring the parameters U, W, G, and S at the midpoint $(x_{i-\frac{1}{2}}, z_{n-\frac{1}{2}}, \eta_{j-\frac{1}{2}})$, the resulting difference form is indicated on the panel. As can be seen, the finite difference form of a parameter is defined in terms of its four values the corners of the box at a given j-point. Note that the three values $q_j^{i-1,n}$ $q_j^{i,n-1}$ and $q_j^{i-1,n-1}$ are known and the value of $q_j^{i,n}$ is unknown.

A different centring arrangement is necessary for e_j, which is related to the normal velocity component, in order to avoid oscillations in the solution procedure. Thus e_j is written in the form shown in the panel with $\bar{e}_j$, the unknown, expressed in terms of its values at all four corners.

13.12 Characteristic Box

When negative spanwise velocities are encountered, write equations (1c), (1d) of panel 13.10 as

$$(bG)' + eG - m_2 U^2 - m_5 UW - m_8 W^2 + m_{11} c = \lambda \frac{\partial U}{\partial s} \qquad (3a)$$

$$(bS)' + eS - m_4 UW - m_3 W^2 - m_9 U^2 + m_{12} c = \lambda \frac{\partial W}{\partial s} \qquad (3b)$$

where

$$\lambda = [(m_{10} U)^2 + (m_7 W)^2]^{1/2}$$

and s denotes the local streamline as shown in the figure

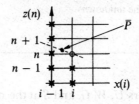

When negative spanwise velocities, i.e. $W_{j-\frac{1}{2}}^{i,n} < 0$, are encountered, the Characteristic Box method introduced by Cebeci and Stewartson (1978) is required. It is based on the solution of the governing equations along local streamlines. Again, we use the procedure of panel 13.10 for equations (1a) and (1b) and centre them at $(x_i, z_n, \eta_{j-\frac{1}{2}})$. The finite difference approximations to equations (1c) and (1d) are first written in the streamline form shown in the panel.

Since the local streamline direction is known for the initial U and W profiles, we centre equations (3) at $\bar{P}$ as shown and write the finite difference approximations as usual. To solve the continuity equation (equation (1e) of panel 13.7), however, we use the zig-zag scheme discussed on panels 8.6 and 8.7 and centre it as before. When $W_{j-\frac{1}{2}}^{i,n} > 0$, we revert to the standard Box scheme and centre the equations in the manner described in panel 13.11.

13.13 Results for transonic flows

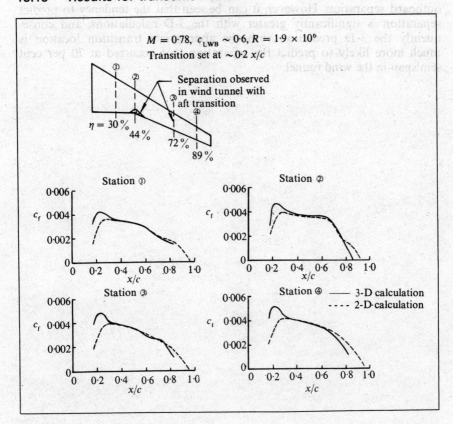

The Cebeci, Kaups and Ramsey method discussed in the previous panels has been applied to subsonic and transonic flows and the calculated results for typical large transport wings are reported by Lynch (1978). The above figure shows the computed chordwise c_f distribution at several spanwise stations for a Douglas wing at 0·78 Mach number together with those obtained by using two-dimensional strip theory approximation. Transition is set at about 20 per cent chord for both calculations. The predictions with this transition location indicate separation would occur all across the span, being most critical at the trailing-edge break station. However, the predicted separation is eliminated all across the span when the further-aft wind tunnel-test transition location is used in the calculation. With the 3-D boundary-layer calculations, the separation at the trailing-edge break station is predicted to occur further forward, and hence is not likely to be eliminated with the further-aft transition location. When the 3-D calculations predicted separation at about 85 per cent chord point at the trailing-edge break station, the calculations did not continue outboard. Therefore, there is no prediction of any outboard separation. However, it can be seen that the tendency to predict separation is significantly greater with the 3-D calculations, and consequently the 3-D prediction with the appropriate transition location is much more likely to predict the separation that occurred at 70 per cent semispan in the wind tunnel.

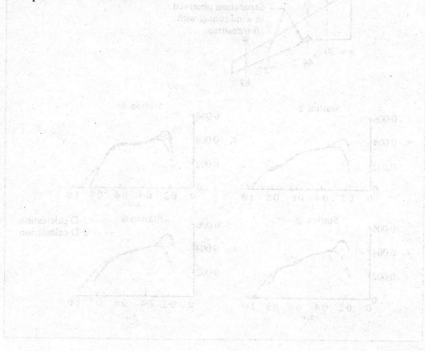

13.14 Inviscid–viscous interaction for a three-dimensional flow

To compute *displacement surface*, solve

$$\frac{\partial}{\partial x}|U_e h_2 \sin\theta(\Delta - \delta_x^*)| + \frac{\partial}{\partial z}|W_e h_1 \sin\theta(\Delta - \delta_z^*)| = 0 \qquad (1)$$

by expressing it as two ordinary differential equations

$$\frac{dz}{dx} = \frac{W_e h_1}{U_e h_2}, \qquad \frac{d\Delta}{dx} = \frac{f(x,z)}{U_e h_2 \sin\theta} \qquad (2)$$

Here

$$f(x,z) = \frac{\partial}{\partial x}(U_e h_2 \sin\theta \delta_x^*) + \frac{\partial}{\partial z}(W_e h_1 \sin\theta \delta_z^*) - \Delta\left[\frac{\partial}{\partial x}(U_e h_2 \sin\theta)\right.$$
$$\left. + \frac{\partial}{\partial z}(W_e h_1 \sin\theta)\right] \qquad (3)$$

Compute V_w by

$$V_w = \frac{1}{h_1 h_2 \sin\theta}\left[\frac{\partial}{\partial x}(U_e h_2 \sin\theta \delta_x^*) + \frac{\partial}{\partial z}(W_e h_1 \sin\delta_z^*)\right] \qquad (4)$$

To account for the effects of viscous forces on the pressure distribution, we must allow for the inviscid–viscous interaction. This can be done in two ways. One way is to compute the displacement surface Δ for a given inviscid pressure distribution, calculate the potential flow about this modified shape and iterate until some convergence criterion is satisfied.

In order to compute the displacement surface for incompressible flows, it is necessary to solve the first-order equation (1) given in the above panel. Here x and z are two surface coordinates; θ is the angle between the coordinate lines; h_1, U_e and h_2, W_e are the metric coefficients and inviscid velocity components associated with the x and z coordinate directions, respectively; δ_x^* and δ_z^* are the displacement thicknesses

$$\delta_x^* = \int_0^\infty \left(1 - \frac{U}{U_e}\right) dy, \qquad \delta_z^* = \int_0^\infty \left(1 - \frac{W}{W_e}\right) dy. \qquad (4)$$

Note that the first equation in (2) is for the calculation of the inviscid streamlines and can be solved independently of the second one. The solution for Δ then proceeds by integration along the individual streamlines, starting with assumed or calculated values near the stagnation line of the wing.

The second way is to simulate the viscous effects by the blowing approach in which we distribute sources on the body. The blowing velocity V_w, which is given by equation (4) in the panel, is used as a boundary condition in the potential flow solution.

13.15 Results for inviscid–viscous interaction on a wing

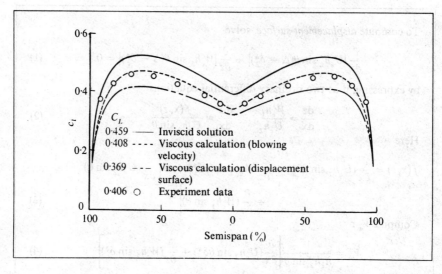

The blowing approach for subsonic flows has the advantage that the original body shape does not change during calculations and thus that the potential flow influence matrix and its inverse can be saved. The result is that all potential-flow calculations for the various iterations are accomplished in the same computing time as a single potential flow calculation. Though this approach has computational advantages, both approaches must be explored for consistency.

The figure in the above panel shows a comparison of the calculated and measured spanwise lift distribution on a relatively low aspect ratio wing tested by Kolbe and Boltz (1951). The boundary-layer calculations were made by the Cebeci, Kaups and Ramsey method and the inviscid flow calculations by the Hess method. The wing sweep is 45 degrees and the Reynolds number for the test condition is given as 18×10^6. The viscous effects, amounting to a loss of 12 per cent in lift, are evidently predicted well by the blowing approach but not by the displacement surface approach which results typically in over-correction. Comparison with calculations based on strip theory and two-dimensional boundary-layer theory shows that the blowing approach alters the original inviscid solution by small amounts and the displacement-surface addition consistently underpredicts the corrections. The essential difference between calculations based on two- and three-dimensional boundary-layer methods is that the crossflow calculated by the latter increases both the displacement surface height and the blowing velocity near the trailing edge of the wing. The reason for the difference in results obtained by the displacement surface and blowing approaches in three dimensions is not known with certainty but it is very likely that it stems from the omission of the wake effect and its relationship to the conditions imposed at the trailing edge.

In viewing the above results, it should be noted that the calculations were terminated after the first iteration cycle and that the transition location on the lower surface was not recorded during the experiments. Since no reliable method exists for the prediction of transition in three-dimensional flows, some variation in results can be produced by assuming plausible but different transition locations. In addition, the results are not entirely independent of the potential flow solution since a comparison of solutions obtained with 8 and 20 lifting strips showed that, although the predicted spanwise lift distributions were identical, the magnitude of the individual inviscid velocity components was different and resulted in different boundary-layer characteristics. Obviously further studies are needed.

13.16 Concluding remarks

The procedures described in the preceding pages of this chapter have been successfully applied to a calculation of the flow around wings and also, with some modifications to the boundary conditions, to the flow around ship hulls: further details of these two applications have been reported by Cebeci Kaups and Ramsey (1977) and Cebeci, Chang and Kaups (1980), respectively. It is probably the most advanced and practically useful procedure available at the present time; improvements are still possible and are under investigation. Nevertheless, the procedure can be used now to provide information of relevance to design.

14 Turbomachinery

14.1 Purposes and outline of chapter

- Axial flow (airfoil-like blades): compressors and turbines
- Centrifugal flow (one rotor and diffuser): compressors only
- Special turbulence problems: streamline curvature effects (axial machine blades) or Coriolis forces (centrifugal impeller), and imbedded longitudinal vortices (in passage corners or on annulus walls)

The two basic geometries for air or gas turbomachines are axial flow, with alternate fixed and rotating rows of highly cambered airfoil-like blades, and centrifugal flow (for compressors only) with an axial intake to a multiple-vane impeller (rotor) and radial outlet to a vaned or vaneless diffuser. Radial-flow water turbines are common but will not be discussed here.

Detailed calculations of centrifugal machine flows are rare but a variety of methods has been developed for calculation of parts of axial machine flows. Axial machine blades have aspect ratios between about 1 and 5, and flow in the blade–annulus junctions (leading to very complicated flow on the annulus walls) has a large effect on performace. To date the crudity of the turbulence models used has reflected ignorance of the behaviour of turbulence in these flows.

The basic problems in axial or centrifugal machines are the effect of skew-induced trailing mean vorticity, and Coriolis or "centrifugal" (streamline curvature) forces. These have been discussed, in general terms in previous chapters. This chapter is an introduction to turbomachine problems (for non-experts) and a commentary on ways of closing the gap between the simple turbulence models in current use by turbomachine designers and the more refined ones described in previous chapters.

Panels 14.2 and 14.3 deal with turbomachine geometries, emphasizing the extreme complications of the flow in centrifugal compressors, and panel 14.4 discusses machine efficiency; axial compressor or turbine efficiencies are already over 90 per cent but aerodynamic improvement can also lead to reductions in size and weight of the machine. Panel 14.5 presents the main difference between the environments of wings and of turbomachine surfaces, the high free stream "turbulence" experienced by the latter. Except near root and tip, axial machine blade boundary layers are only mildly three-dimensional; thus, if adequate allowances can be made for free-stream turbulence, high surface curvature and the effects of transition and low Reynolds number, standard calculation methods can be used (panel 14.6). Panels 14.7 to 14.9 discuss blade wakes and their contribution to the free-stream disturbances encountered by the downstream blade rows. Panels 14.10 to 14.12 treat annulus (casing) boundary layers, which are circumferentially periodic and are highly disturbed by the vortices shed from the blade roots and tips; usually rather crude integral methods are employed. Annulus flows are the main region in which more refined method can make an impact but the imbedded vortices form a severe test of any turbulence model (Chapter 11): for a comparison with calculations by two advanced methods, see Higuchi and Rubesin (1978).

Turbomachinery

14.2 Basic geometries, 1

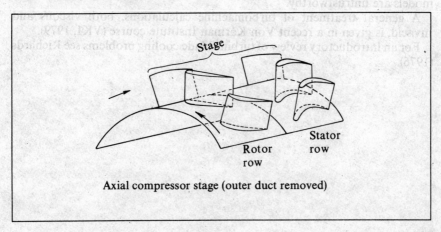

Axial compressor stage (outer duct removed)

In axial flow machines the main problem areas are the boundary layers and wakes of the blades, and the boundary layers on the inner and outer walls of the annulus (the "end walls" of the blade passages).

Three difficulties inseparable from the rotation of the machine are

(i) the abrupt change in hub surface velocity at the gap between the rotor row and the stator row of each stage;
(ii) the leakage through the gap between the rotor-blade tips and the duct wall, combined with the relative motion of blade and wall;
(iii) the periodic encounters between blades and the wakes of the blades in the row upstream.

The first has been studied experimentally by Bissonnette and Mellor (1974) and by Lohmann (1976). The second is usually dealt with empirically (the main effect being a complicated form of tip vortex). The third difficulty has often been dealt with in the past by regarding the wakes as part of the free-stream turbulence.

The annulus boundary layers are greatly affected by skew-induced vorticity generated at the blade leading edges (see Chapter 11).

The pressure rises through compressors and falls through turbines. Separation is therefore not a serious problem in the latter, heat transfer is. Flow over turbine blades is often laminar, despite high free-stream turbulence. Blade Reynolds numbers in compressors or turbines are in the range 10^5–10^6 (or more for large steam turbines) so transition and/or low Reynolds number effects are important.

A short review of turbulence problems in axial machines is given by Johnston (Bradshaw, 1976, Chapter 3). Basic calculation methods for flow over blades are similar to those for flow over airfoils, but annulus boundary layers are so disturbed by streamwise vorticity that conventional turbulence models are untrustworthy.

A general treatment of turbomachine calculations, both viscous and inviscid, is given in a recent Von Kármán Institute course (VKI, 1979).

For an introductory review of turbine blade cooling problems see Richards (1976).

14.3 Basic geometries, 2

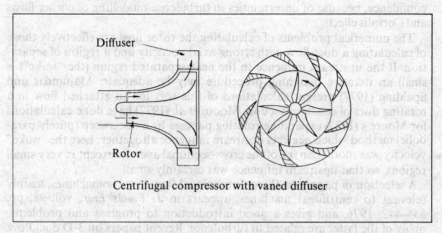

Centrifugal compressor with vaned diffuser

"Squirrel cage" blowers are used only for small sizes and small pressure rises. All centrifugal machines with pressure ratios greater than 1·1 are of the form shown except that some have vaneless diffusers in which the part of the kinetic energy due to swirl is not recovered.

Because of the long passages and complicated blade twist, boundary layers in the radial part of the impeller (rotor) are thick, generally near separation on the trailing (suction) side, and subject to strong secondary flow. Coriolis forces have a large effect (Moore, 1973; see also Bradshaw, 1973, for further references).

The flow entering the diffuser is usually supersonic relative to the vanes. Rotor tip speed (and therefore the output of a given size of machine) is limited by the loss in efficiency due to shock–boundary-layer interaction in the diffuser. Speeds in excess of 2500 ft s^{-1} (750 m s^{-1}) have been used.

Mainly because of losses caused by shock–boundary-layer interaction, the efficiency of the best machines falls from about 85 per cent for pressure ratios of roughly 2 to 75 per cent or less for machines designed for pressure ratios of 10 or more. Because of the large separated region in the impeller the radial velocity of the flow reaching the diffuser entry fluctuates greatly and calculations based on mean velocities relative to the diffuser may be unrealistic. The diffuser vanes are usually thick with blunt bases as shown. The alternative would be to use thin vanes with a contraction in passage height to give an acceptably small effective divergence angle: a dump diffuser with circumferential backward-facing steps would then be needed to match the diffuser proper with the annular collector.

Research knowledge of the flow in centrifugal machines is limited: development has so far proceeded by well-informed trial and error rather than by detailed calculations (e.g. Lakshminarayana *et al.*, 1974, and R. C. Dean, 1974). The work of Eckardt (1975) has greatly improved our knowledge of the flow out of a rotor. It is undoubtedly too complicated to calculate with confidence, because of uncertainties in turbulence modelling of corner flows and Coriolis effects.

The numerical problems of calculating the rotor flow are effectively those of calculating a duct flow with strong axial vorticity and a region of separation. If the upstream influence in the near-separated region (the "wake") is small an iterative marching procedure may be adequate. Majumdar and Spalding (1977) present calculations of this sort for an attached flow in a rotating duct of constant section. Moore *et al.* (1977) have done calculations for Moore's (1973) expanding rotating passage by a one-sweep (purely parabolic) method which neglects upstream influence altogether: here the "wake" velocity was more than 0·1 of the cross-sectional average except in very small regions, so that upstream influence was certainly small.

A selection of papers on three-dimensional flow in turbomachines, mainly relevant to centrifugal machines, appears in *J. Fluids Eng.*, vol. 98, pp. 354–442, 1976, and gives a good introduction to progress and problems; many of the latter are related to turbulence. Recent papers on 3-D duct flow, with emphasis on numerical methods, include Eiseman *et al.* (1978) and Roberts and Forester (1979). A key question is whether upstream influence (panel 11.6) is significant. If not, a parabolic solution like Roberts and Forester's can be used; else, an elliptic method (Chapter 10) is needed.

14.4 Improvements in efficiency

- Compressor efficiency = (isentropic T_0 rise)/(actual T_0 rise) >0.9
- Propulsive efficiency $\simeq$ engine $\eta \times 2U/(U + U_j)$—improvable by reduction in U_j (increased bypass ratio) or improvement in engine η
- Improved design can also decrease component size, weight or cost

Although the component efficiency of gas turbines is high (axial compressor efficiencies are over 90 per cent) overall propulsive efficiency (inversely proportional to specific fuel consumption) can still be improved. Rolls-Royce hope for 20 per cent reduction in sfc by further development of the RB 211 (whose first model had 20–25 per cent lower sfc than straight-through engines). Most of this reduction will come from an increase in bypass ratio (and therefore a decrease in bulk-average jet velocity) which brings aerodynamic problems of its own, but improvements in component efficiency are expected. Improved design can also lead to *smaller* components as well as better energy efficiency.

Combustion is dealt with in Chapter 15: here we treat compressors and turbines.

14.5 Free-stream "turbulence"

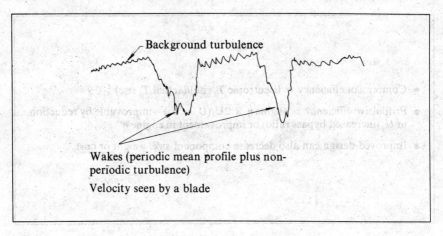

In the latter stages of an axial flow compressor the free-stream "turbulence" has reached a near-constant level of the order of 5 per cent of the relative velocity. The fluctuations are composed partly of the well-mixed wakes of blade rows far upstream, which count as true (though perhaps not isotropic) turbulence, and partly of the discrete wakes of the last one or two rows. These blade wakes carry their own turbulence, whose length scale is necessarily rather larger than that of the boundary layer on the blades that encounter the wakes; because of the relative motion of rotor and stator the mean velocity profile of the wake appears as periodic pulses.

For design purposes it is important to distinguish (i) true turbulence of a scale likely to affect the boundary layer, (ii) larger-scale nonperiodic unsteadiness, (iii) ordered unsteadiness. The last-named includes not only the wakes as such but also the "inviscid" variations of speed and direction of the flow outside the wakes, which may be significant near the front of the following blades if the inter-row gap is small.

At present, only rather simple empirical correlations, which frequently lump (i)–(iii) together as "percentage turbulence" $\sqrt{\overline{u^2}}/U$, are in common use in turbomachine calculation methods.

Turbomachinery 281

14.6 Blade boundary layers

- High free-stream turbulence and ordered unsteadiness
- Large surface curvature
- Low Reynolds number
- Transpiration cooling (turbines)
- Three-dimensional and Coriolis effects

Present knowledge of free-stream turbulence (Bradshaw 1976) suggests that eddies whose wavelength is of order δ increase boundary-layer skin friction by 2 to 4 per cent for one per cent rms velocity fluctuation, and increase heat transfer slightly more. The direct effects are on the outer layer. Calculation methods which are empirically adjusted to reproduce the increase in flat-plate skin friction should be adequate in arbitrary pressure gradient.

Larger-scale unsteadiness which does not directly affect turbulence structure presents no difficulty to transport equation methods. D/Dt is still defined as $\partial/\partial t + U_i \partial/\partial x_i$ (where the averages are now taken over at least one cycle of the unsteadiness) and the relative size of temporal and spatial derivatives is unimportant as long as D/Dt is not too large.

Surface curvature effects on turbulence, discussed in Chapter 1, are important in all types of turbomachine. It is well known that concave curvature hastens transition, by Taylor–Görtler instability. Convex curvature does not affect two-dimensional (Tollmien-Schlichting) instability but probably affects the later (three-dimensional) stages of transition.

Apart from injection cooling, for which see the review and experimental work of Choe *et al.* (1975) and Crawford *et al.* (1976), blade boundary layer problems are close to those found on wings or in duct flows and no dedicated calculation methods are needed. Cooling is effected by injection through arrays of holes whose diameter may be several times δ. Skew-induced streamwise vorticity (panel 11.15) may *increase* heat transfer near the hole. Cooling can be predicted either by data correlations or by a detailed calculation of the jet–boundary-layer interaction. Upstream influence of the jet may be significant near the injection hole so that equations elliptic in three dimensions must be solved: see Chapter 10.

Three-dimensionality in blade boundary layers results from (i) the small component of Coriolis force tangential to the blade surface, (ii) spanwise variations in pressure distribution due to changes in blade loading, (iii) crossflow originating near root and tip. Frequently all are neglected.

14.7 Blade trailing edges

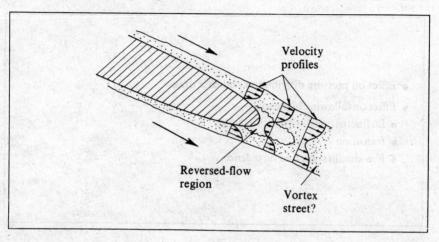

Turbomachine blade chords are typically 3 to 10 cm and trailing-edge radii may be 1 per cent of chord or more. To determine the circulation round the blade the separated flow region must be calculated or approximated. Geller (see Surugue, 1972) shows that adequate results can be obtained by free-streamline theory, assuming a stagnant wake extending to infinity. A full solution could be obtained (Chapter 10) assuming the turbulence model to be adequate: evidently it is not critical for determining the circulation but a fairly accurate calculation of the base pressure is needed to give the total drag of the blade and the momentum deficit in the wake.

An alternative to transpiration cooling of turbine blades is internal cooling by air which is then ejected through the base (probably through elongated holes rather than a near-continuous slot). This further increases the trailing edge thickness needed for strength but may give a net reduction in base drag to offset the cost of pumping the cooling air. Note that Geller's model applies to this case also.

Lawaczeck *et al.* (1976) have shown that the wake behind a turbomachine blade with a finite trailing-edge radius can develop into a vortex street—that is, a feedback-supported oscillation of the blade circulation can occur. The resulting fluctuating load on the blade may be important, and so may the effect of the vortex streets on the next row of blades.

14.8 Blade wakes

- Effect on pressure distribution and base drag
- Effect on following blade row—
  - lift fluctuations → vibration, noise
  - transition
  - free-stream shear and turbulence

Blade wakes affect the pressure distribution over the blade, just as in the case of airfoils: indeed the effects are larger because of the large trailing-edge radius needed for structural integrity. However the main effect of blade wakes is on the blades in the next row downstream, which intersect them periodically. Near the nose the blade boundary layer is so thin that the effect of the wake acts via the inviscid flow: the inter-blade spacing is determined partly by the structure of the blade root and partly by the need to avoid fluctuating loads on the downstream blade. The minimum velocity in the wake, measured relative to the blade causing the wake, rises rapidly to about 0·6–0·7 of the free-stream speed (i.e. the velocity at the edge of the inner layer) but thereafter rises very slowly. Excessive inter-row spacings would be needed to make periodic loading negligible.

The particularly disturbing propeller-like noise emitted by some turbofan engines is related to the encounters of the downstream stator blades with the wakes of the fan blades. The noise is at the *shaft* frequency and its first few harmonics and is due primarily to small differences between blades at different parts of the circumference. The noise at the *blade* passing frequency is attenuated in the duct.

14.9 Blade wake interactions

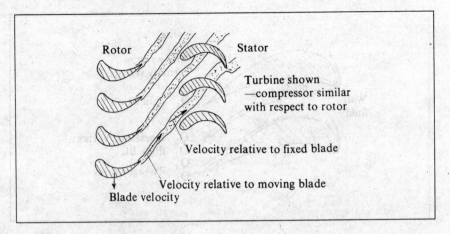

Wakes usually intersect the following blade rows at a large angle, because of relative motion. A good qualitative description is given by Walker and Oliver (1972) and Walker (1974). Discussion in terms of shear-layer interaction (panel 1.16) is not appropriate because of the large angle of intersection, and for calculation purposes it may be adequate to consider only the mean shear in the wake and not its turbulence.

Since the wake width considerably exceeds the boundary layer thickness, the interaction can be considered as

(i) quasi-inviscid interaction with pressure field of blade; spanwise component of vorticity is conserved, normal component of velocity is virtually destroyed;

(ii) effect of fluctuation in U_e and $\partial U/\partial y$ on blade boundary layer.

Part (i) is a difficult unsteady-flow problem, outside the scope of this book. Passage of the wake over the trailing edge may lead to significant changes in circulation.

Part (ii) is an unsteady boundary layer problem: response to U_e fluctuations is discussed in Chapter 8. Experiments on boundary layers with *steady* non-zero $\partial U/\partial y$, in the presence of some free-stream turbulence, are reported by Ahmad *et al.* (1976). The effects of $\partial U/\partial y$ are relatively small and confined to the outer layer: from the point of view of predicting the *mean* growth and skin friction in the blade boundary layer it may be permissible to ignore the wake shear, and possibly the effect of the large-scale quasi-two-dimensional vortices if the wake contains a vortex street (panel 13.7). It is unlikely that the blade–wake interaction will be strong enough to induce separation in the interaction region itself as this would imply unacceptably large blade vibration.

14.10 Annulus boundary layers

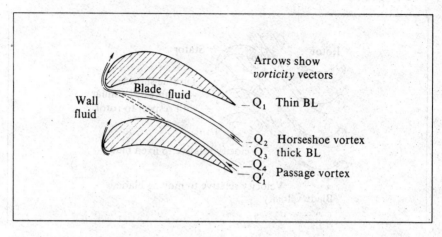

Useful reading for those wishing to calculate annulus boundary layers includes the flow visualization study by Herzig et al. (1953) and the combined visualization and quantitative measurements of Langston et al. (1977). Herzig et al. used mostly very thin blades so that the horseshoe vortices wrapped round the blade leading edges in the junction region were weak. Langston et al. used thick blades in a turbine cascade with a turning angle of 110° and a velocity ratio of 1·6. The horseshoe vortex extends right across the passage (as shown dotted above) and most of the fluid in the original end wall boundary layer is dumped into the passage vortex on the upper (suction) surface of the blade below the one that generates the horseshoe vortex. The new end-wall boundary layer consists of fluid from the (thin) boundary layer on the pressure surface of the blade. The "boundary layer" entering the next blade row consists of alternate passage vortices and thin, possibly laminar, strips of boundary layer. In less extreme cases the horseshoe vortex will move only partway across the passage (full lines above) and the emerging wall "boundary layer" will have

    a. a thin boundary layer of blade fluid from Q_1 to Q_2;
    b. the counterclockwise leg of the horseshoe vortex from Q_2 to Q_3;
    c. end-wall fluid (thicker boundary layer) from Q_3 to Q_4.

The clockwise leg of the horseshoe vortex from Q_4 to Q'_1 will usually have migrated on to the blade, leaving an anticlockwise "passage vortex" of skewed wall fluid. Further downstream the vortices will further thicken the boundary layer between Q_3 and Q_4.

The older calculation methods (see panel 13.12) are generally based on passage-averaged equations integrated across the gap between the blades (and frequently integrated across the boundary layer thickness, leading to ordinary differential equations).

Turbomachinery

Some attempts have been made to use eddy-viscosity-transport methods in curved turbomachine passages (e.g. Majumdar and Spalding, 1976; see also Chapter 10 on curved-duct flows in general) but it is doubtful whether an isotropic eddy viscosity is adequate in such cases.

14.11 Effect of tip and shroud gaps

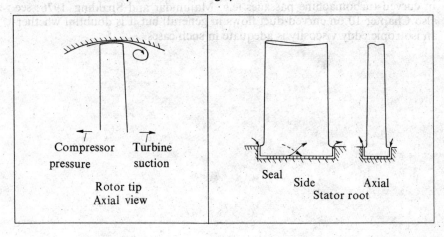

At the rotor tip, fluid leaks from the pressure side to the suction side, tending to form a vortex in the same sense as a wing trailing vortex but the opposite sense to the passage vortex formed by skewing, and cross-stream migration, of end-wall fluid. Generally the two vortices remain distinct. Usual design procedure is to allow an empirical factor for tip losses. The relative motion of the duct wall resists the tip clearance flow in a turbine and assists it in a compressor.

Stator blade rows are usually built into the annulus with shrouds and seals at the hub. Fluid can leak from front to back of the blade under the seals: it emerges with negligible momentum and therefore adds to the thickness of the hub boundary layer without inducing vorticity.

It seems unlikely that either of these effects would be simulated in detail in a calculation method, if only because of excessive demands on finite difference resolution.

14.12 Integral methods for annulus boundary layers

- Take an integral method for $\partial/\partial x$ and $\partial/\partial z$ of profile parameters. Average with respect to circumferential distance, z, between adjacent blades. Add x and z components of "blade force" to represent
  i Δp across blade thickness
  ii nonlinear pseudo-stress terms
  iii effect of corner boundary layers and vortices

- "Blade force" may vary with y so an integral over y is needed. Deficit quantities must be evaluated accurately

---

Integral methods for two-dimensional shear layer calculation can be derived either by the generalized Galerkin technique of weighted integration, with respect to y, of equations, such as those of the panel, or by control-volume analysis. In either case the result is a set of ordinary differential equations in x whose unknowns are profile parameters (θ, H, c_f). In 3-D, integral methods consist of partial differential equations in the surface coordinates x and z.

Horlock and Perkins (1974) and Balsa and Mellor (1975) have developed methods in which the equations for an annulus wall boundary layer are averaged circumferentially across the blade passage leaving ODEs in the meridional (near-axial) coordinate x. Terms resulting from this averaging of nonlinear equations are lumped with the "blade force" applied at the circumferential ends of the averaging zone. It has been found that the blade force varies through the boundary layer: an empirical expression is needed for its integral from 0 to δ. The assumptions made for the development of H are quite simple—Balsa and Mellor use constant H, while Horlock and Perkins use Head's entrainment assumption. Balsa and Mellor point out that care is needed in formulating "deficit" quantities if different approximations are made in the external stream and in the boundary layer. Some of the basic ideas are due to Dring (1971).

These are attempts to get easy answers to a difficult problem. Obviously, pitch-averaged equations are acceptable only if the variations across the passage are small: see Sockol (1978) for a review.

Use of a standard integral method for 3-D boundary layers, without averaging across the passage, still involves an estimate of the "blade-force", which has to simulate the effects of the shear layers in the blade–annulus junction, where boundary-layer equations and turbulence models do not apply.

A prerequisite for a rational solution is an understanding of turbulence in streamwise corners and imbedded vortices (Chapter 11).

15 Combustion

15.1 Purposes and outline of chapter

- To introduce physical and chemical features of combusting flows and their implications for related calculations
- To discuss equations appropriate to turbulent, combusting flows and, in particular to introduce mass weighted averaging and its possible benefits for calculation purposes
- To present combustion models for diffusion, premixed and partially premixed flames and to indicate relative advantages

It will be clear, from preceding chapters, that the calculation of turbulence-controlled flows can involve complicated equations and numerical schemes. Combusting flows introduce the added complications of turbulence-chemistry interactions plus high-temperature heat-transfer processes and their calculation requires, therefore, additional assumptions. The precision of calculated flow properties in combusting flows must be expected to be worse than those for isothermal flows in similar geometries. The need to improve the design of furnace and combustor configurations is, however, great and has been increased by the current trend towards energy conservation and pollution control. As a result, considerable effort is presently being devoted to the development of calculation methods for combusting flows even though their accuracy is known to be limited.

The pressure for the development of calculation methods is related to flows confined within complex geometries and, in most cases, these are represented by elliptic equations in two or three dimensions. A major purpose of the related numerical procedure is to determine the influence of changes in geometric features or in fuelling arrangements on local flow and surface properties. Thus the absolute values of the calculated properties are of lesser significance and the accuracy of calculation may be tolerable. The development of combustion models is best achieved with simple flows, such as those represented by two-dimensional parabolic equations.

15.2 Physical and chemical features of flames

> - The fuel may be gaseous, liquid or solid and may influence the flow even without chemical reaction
> - Stabilization is necessary
> - Chemical reaction takes place locally and, as well as increasing temperature, will reduce density
> - Measurements of temperature fluctuations, though imprecise, indicate very large values
> - Chemical reactions can be fast or slow, relative to turbulent diffusion
> - Radiation can be important

The nature of the fuel, and the way in which it is provided, is of great importance. In unconfined gaseous diffusion flames, the flow may resemble a jet with the luminous region providing a visual impression. In this case, and provided the initial velocity is large enough to ensure turbulent flow in spite of the reduction in Reynolds number caused by the increase in temperature, a hot turbulent jet-like flow will occur and will be turbulence controlled. The temperature fluctuations will be larger than those associated with non-combusting jet flows and the effect of radiation will be small. In a corresponding jet flame with premixed fuel and air, chemical characteristics control the combustion in some regions but turbulence characteristics will still be a major controlling influence.

With liquid or solid fuels, the initial trajectory of the fuel will persist for a distance which will depend on the size of the droplets or particles. It is possible that the structure of the surrounding gaseous phase will be influenced by the fuel and the temperature distribution, for example, will depend on the location and rate of burning. Radiation is likely to be important.

Confined flames, such as those of furnaces and combustors, are usually required to transfer heat by radiation, may be partially premixed and involve reactions which are sufficiently slow to exert some control. In general, however, the turbulence-chemistry interactions will be very important.

A corresponding calculation method will involve models for turbulence, combustion and radiation. The first two are considered on the following panels. Radiation can be treated separately and the reader is referred to, for example, Lockwood and Shah (1978) for further information.

15.3 Equations and averaging, 1

- Unweighted: $\hat{U}_i = \overline{U}_i + u'_i, \qquad \hat{\Phi} = \overline{\Phi} + \Phi'$
  $$\overline{u'_i} = 0, \qquad \overline{\phi'} = 0$$

- Density weighted: $\hat{U}_i = \tilde{U}_i + u''_i, \qquad \hat{\Phi} = \tilde{\Phi} + \phi''$
  $$\overline{\rho u''} = 0, \qquad \overline{\rho \phi''} = 0$$
  $$\overline{u''_i} \neq 0, \qquad \overline{\phi''} \neq 0$$

where overbar denotes ensemble averages, i.e.
$$\overline{U}_i = \frac{1}{N} \lim \sum_{n=1}^{N} (\hat{U}_i)_n$$

The use of unweighted ensemble averages, similar to those used in previous chapters, will result in correlations between the fluctuating density and other flow properties. These can be ignored or determined from further modelled equations. As will be shown, the neglect of density-fluctuation correlations leads to results which are significantly different from those obtained when they are included. Undoubtedly, however, the assumptions inherent in the modelled equations for $\overline{\rho' u_i}$ and $\overline{\rho' \phi'}$ lead to uncertainties and it is difficult to judge which is more correct.

A further uncertainty is introduced when comparison with measurements are made. Velocity and concentration measurements, by sampling probes, are close to the density-weighted average, i.e. $\tilde{\phi}$. Temperature is less easy to specify; with large thermocouples and pyrometers measurements are probably closer to the density-weighted temperature but with small probe dimensions (e.g. < 40 μm) can be close to the unweighted average.*

---

* In contrast to the notation of panel 2.5, and for reasons of symmetry, overbars are used here for ensemble-averaged quantities.

15.4 Equations and averaging, 2

- For high turbulence Reynolds numbers, the momentum equation with unweighted averages has the form

$$\frac{\partial}{\partial t}[\bar{\rho}\bar{U}_i + \overline{\rho'u'_i}] + \frac{\partial}{\partial x_j}[\bar{\rho}\bar{U}_i\bar{U}_j + \bar{\rho}\overline{u'_iu'_j} + \bar{U}_j\overline{\rho'u'_i}]$$
$$+ \frac{\partial \bar{P}}{\partial x_j} - \bar{\rho}g_i = 0$$

- With density-weighted averaging, as proposed by Favre (1965), the equation has the form

$$\bar{\rho}\frac{\partial \tilde{U}_i}{\partial t} + \bar{\rho}\tilde{U}_j\frac{\partial \tilde{U}_i}{\partial x_j} + \frac{\partial \bar{P}}{\partial x_i} + \frac{\partial}{\partial x_j}(\bar{\rho}\widetilde{u''_iu''_j}) - \bar{\rho}g_i = 0$$

The panel shows the momentum equation in the forms which stem from conventional and density-weighted averaging. As can be seen, the density-weighted equation is identical to that considered previously for isothermal flows with the single exception that density weighting overbars are present. In contrast, the ensemble-averaged equation contains the additional density-fluctuation correlation previously referred to. These correlations can be regarded in the same way as the components of the stress strain tensor and conservation equations written with them as dependent variables. The number of $\rho'\phi'$ correlations is likely to be large to represent a reacting flow and assumptions, similar to those already described, will be required in all equations.

In the use of the density-weighted averaged equations, Favre averaged (1965), it is useful to note that a turbulence model is now required to represent the quantity $\widetilde{u''_iu''_j}$. This is discussed in relation to the following panel.

The link between the two equations, in terms of the averages, may be expressed by the relationships:

$$\bar{U}_i = \tilde{U}_i + \overline{u''_i},$$
$$\overline{u''_i} = -\overline{\rho'u_i}/\bar{\rho},$$
$$\overline{\rho'u''_i} = \overline{\rho'u_i},$$
$$\overline{u'_ju'_i} = \widetilde{u''_iu''_j} - \overline{\rho'u''_iu''_j}/\bar{\rho} + \overline{u''_i}\,\overline{u''_j}$$

15.5 Equations and averaging, 3

In density-averaged form, the kinetic energy and dissipation-rate equations are

$$\bar{\rho}\frac{\partial k}{\partial t} + \bar{\rho}\tilde{U}_i\frac{\partial k}{\partial x_i} + \overline{\bar{\rho}u_i'' u_j''}\frac{\partial \tilde{U}_i}{\partial x_j} - \frac{\partial}{\partial x_j}\left[\frac{\mu_t}{\sigma_k}\frac{\partial k}{\partial x_j}\right] - \frac{\overline{\rho' u_i''}}{\bar{\rho}}\frac{\partial \bar{P}}{\partial x_i} + \bar{\rho}\varepsilon = 0$$

$$\bar{\rho}\frac{\partial \varepsilon}{\partial t} + \bar{\rho}\tilde{U}_i\frac{\partial \varepsilon}{\partial x_i} + C_{\varepsilon,1}\bar{\rho}\frac{\varepsilon}{k}\overline{\tilde{u}_i'' u_j''}\frac{\partial \tilde{U}_i}{\partial x_j} - C_{\varepsilon,1}\frac{\varepsilon}{k}\frac{\overline{\rho' u_i''}}{\bar{\rho}}\frac{\partial \bar{P}}{\partial x_i} - \frac{\partial}{\partial x_i}\left[\frac{\mu_t}{\sigma_\varepsilon}\frac{\partial \varepsilon}{\partial x_j}\right] + C_{\varepsilon,2}\frac{\bar{\rho}\varepsilon^2}{k} = 0$$

with
$$\mu_t = C_\mu\bar{\rho}k^2/\varepsilon; \quad k \equiv \overline{\tilde{u}_j'' u_j''}/2$$

and
$$\bar{\rho}\,\overline{u_i'' \tilde{u}_j''} = \tfrac{2}{3}\delta_{ij}\left(\bar{\rho}k + \mu_t\frac{\partial \tilde{U}_l}{\partial x_l}\right) - \mu_t\left(\frac{\partial \tilde{U}_i}{\partial x_j} + \frac{\partial \tilde{U}_j}{\partial x_i}\right)$$

A useful discussion of Reynolds stress equations, in density-weighted form, has been given by Jones (1979). Since the number of equations which has to be solved for a combusting flow and the resulting uncertainties are larger than for isothermal flow, there is generally little advantage at present in considering a turbulence model more complicated than the two-equation form indicated on the panel. Further comments can be found in Jones and Whitelaw (1978).

The equations for turbulence energy and dissipation rate, with density averaging, are shown on the panel. They are similar in form to those for the conventional averaged equations for isothermal flow though there is no direct evidence that models based on the two forms of averaging can be treated identically. Certainly, the conventional averaged equations can be written for combusting flows and will contain density–velocity correlations. In the density-weighted form of the turbulence energy equation, the additional term involving $\overline{\rho' u''}$ can be found from the algebraic form of its conservation equation

$$\overline{\rho' u_i''} = -\frac{\mu_t}{\bar{\rho}\sigma_t}\frac{\partial \bar{\rho}}{\partial x_i}.$$

In general, three practical approaches are possible. The first, and most consistent, is to solve the density-averaged equations and presume that turbulence-model assumptions, based on constant density flow, also hold for non-isothermal flow and density-weighted averages. The second is to write the conventional averaged equations with density-fluctuation correlations; solve modelled equations for these correlations; and presume that all assumptions are correct. The third is to ignore density-fluctuation correlations and solve conventional averaged equations. This is the same as the first approach except that the dependent variables are erroneously assumed to be unweighted quantities.

Combustion

15.6 Equations and averaging, 4

- In density-averaged form, the scalar flux equation has the form

$$\bar{\rho}\frac{\partial \widetilde{u_i''\phi_\alpha''}}{\partial t} + \bar{\rho}U_j\frac{\partial \widetilde{u_i''\phi_\alpha''}}{\partial x_j} = -\left[\bar{\rho}\,\widetilde{u_j''\phi_\alpha''}\frac{\partial \tilde{U}_i}{\partial x_j} + \bar{\rho}\,\widetilde{u_i''u_j''}\frac{\partial \tilde{\Phi}_\alpha}{\partial x_j}\right]$$
$$+ \overline{\phi_\alpha''\frac{\partial \bar{P}}{\partial x_i}} + \overline{\phi_\alpha''\frac{\partial p'}{\partial x_i}}$$
$$- \frac{\partial}{\partial x_j}(\bar{\rho}\widetilde{u_i''u_j''\phi''}) + \bar{\rho}\widetilde{u_i''S(\phi)}$$

- Scalar correlation and density-correlation equations have a similar form

The unmodelled form of the velocity-scalar correlation equation, with density weighted averages, is shown on the panel and may be compared with the conventional averaged form of panel 2.11. It can readily be modelled, as indicated for example by Jones (1979). As can easily be appreciated, the modelled equations involve substantial assumptions. If the conventional averaged equations are solved, the density-fluctuation correlation equations will have a related form. Since many such equations can be involved, the number of times the assumptions are made can be considerable.

As a final comment on averaging and turbulence models, a link with probability density distributions is useful. If the probability of a scalar is assumed, as it will be as a necessary part of a useful combustion model, then scalar quantities with both forms of average can be linked explicitly. Thus, if we assume $P(f)$ as a scalar and density-weighted probability distribution, we have

$$\bar{\phi} = \int_0^1 \phi(f)\,P(f)\,df,$$
$$\tilde{\phi} = \bar{\rho}\int_0^1 \frac{\phi(f)}{\rho(f)}P(f)\,df,$$
$$\bar{\rho} = \left[\int_0^1 \frac{P(f)}{\rho(f)}\,df\right]^{-1}$$

15.7 Probability density distributions, 1

The functional dependence of concentration, temperature and density f is non-linear

To evaluate mean values introduce the probability density function for f

[Graph: vertical axis from 0 to 1.0 showing f as a fluctuating trace versus Time]

$p(f)\,df$ represents the fraction of time that f takes a value in the range f to $f + df$

$p(f)$ contains information on all the moments of f, e.g.

$$T = \int_0^\infty T(f)\,p(f)\,df$$

The projection of the time-dependent trace of the signal on to the ordinate of the figure in the panel provides the probability distribution and, as a result, contains all time-averaged information of the dependent variable. Thus, the mean and most probable values and all moments are represented by the distribution; time-dependent information is, however, not represented. The assumption of a specific form of probability distribution implies an $f(t)$ relationship and, if all scalar quantities are assumed to correspond to the same distribution, knowledge of the distribution can lead to complete information of all scalar quantities. If equations for $\bar{f}$ and $\overline{f'^2}$ are solved, an assumed probability–density distribution can be completely specified with the calculated values of $\bar{f}$ and $\overline{f'^2}$ and the property f is completely described. In addition, if other quantities can be linked to f, they are also completely described. In the following pages f should be interpreted as the mixture fraction defined in panel 15.11 and, for example, can be directly linked to temperature if radiation is absent.

The following panel is concerned with three assumptions of probability distributions which have been used. An alternative approach is to attempt to solve an equation with the probability distribution as dependent variable and this possibility has been examined by several authors including Pope (1979). The coalescence–dispersion approach of Pratt (1978) represents a simplified procedure with the same aim. Assumptions can also be made, and equations solved, for joint probability distributions but the application of these more complicated approaches is still some way off.

15.8 Probability density distributions, 2

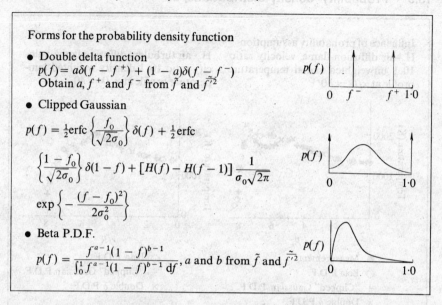

Forms for the probability density function
- Double delta function
  $p(f) = a\delta(f - f^+) + (1 - a)\delta(f - f^-)$
  Obtain a, f^+ and f^- from $\tilde{f}$ and $\widetilde{f'^2}$
- Clipped Gaussian

$$p(f) = \tfrac{1}{2}\mathrm{erfc}\left\{\frac{f_0}{\sqrt{2}\sigma_0}\right\}\delta(f) + \tfrac{1}{2}\mathrm{erfc}$$

$$\left\{\frac{1 - f_0}{\sqrt{2}\sigma_0}\right\}\delta(1 - f) + [H(f) - H(f - 1)]\frac{1}{\sigma_0\sqrt{2\pi}}$$

$$\exp\left\{-\frac{(f - f_0)^2}{2\sigma_0^2}\right\}$$

- Beta P.D.F.

$$p(f) = \frac{f^{a-1}(1 - f)^{b-1}}{\int_0^1 f^{a-1}(1 - f)^{b-1}\,df}, \quad a \text{ and } b \text{ from } \tilde{f} \text{ and } \widetilde{f'^2}$$

Three forms of probability density distribution are shown in the panel and correspond to a rectangular wave form, a random wave form with regimes of $f = 0$ and 1 and to a near random wave represented by an analytic function. The double delta approach was proposed by Spalding and used by Khalil et al. (1975); the clipped Gaussian was proposed by Naguib and Lockwood (1976) and used, for example, by Hutchinson et al. (1977); and the β-function was suggested many years ago and recently used, for example, by Jones (1979) and Jones and Priddin (1978).

In all cases, differential equations are solved for the mean value of a scalar quantity and for the rms of its fluctuations and the resulting two parameters used to fix the distribution. If scalar quantities are presumed to have the same distribution, and corresponding sources are known, they can be determined.

Clearly, the form of a probability density distribution and its influence need to be determined. There are, however, few measurements of probability–density distributions and these are subject to considerable uncertainty. As a result, direct comparisons are usually confined to the resulting temperature distribution. The alternative approach of Pope (1979) may provide guidance as to the most approximate assumptions and could obviate the need for an assumption.

15.9 Probability density distributions, 3

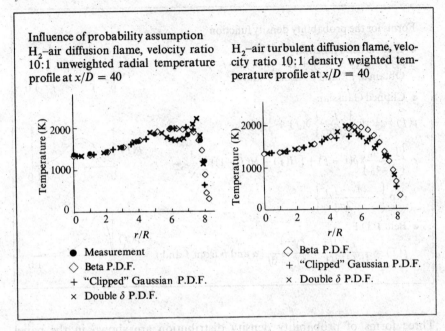

The results of the panel were obtained by Jones (1979) and correspond to the diffusion flame of Kent and Bilger (1972) and to the assumptions of the previous panel. The two figures show conventional and density-weighted averages, respectively, and it should be appreciated that the link between the two is known (see panel 15.6) if the probability distribution is known. The results suggest that the double-delta approach is least satisfactory in that it gives rise to near discontinuities. The maximum difference between the conventional averaged temperature obtained with the Clipped Gaussian and β-function assumptions is less than 100°C and both appear to be in reasonable agreement with the measurements.

Although the measured temperatures have been shown on the left-hand figure it is by no means certain that a thermocouple will measure an ensemble rather than a density-weighted temperature. Indeed, in a later paper, Bilger (1977) has argued that the measurement is close to the density-weighted value but on the opposite side from the unweighted average. This uncertainty implies a limitation on the extent to which combustion models can be appraised and implies the need for improved accuracy of measuring techniques and clear justification before more sophisticated combustion models are used. It should be appreciated, however, that the calculations of the panel are for a hydrogen flame and that the approach used to obtain the results may be less satisfactory with other fuels where intermediate reactions need to be considered.

15.10 Simplest combustion model, 1

Assumptions
- Physical control—infinitely fast, single step chemical reaction
- Mixed is burnt—fuel and oxygen cannot both appear at the same place, even at different times
- Density-fluctuation correlations negligible

Equations

$$\frac{\partial}{\partial x}(\rho U \Phi) + \frac{1}{r}\frac{\partial}{\partial r}(r\rho V \Phi) = S_\Phi + \frac{\partial}{\partial x}\left(\Gamma_\Phi \frac{\partial \Phi}{\partial x}\right) + \frac{1}{r}\frac{\partial}{\partial r}\left(r\Gamma_\Phi \frac{\partial \Phi}{\partial r}\right)$$

$\Phi = U_i, k, \varepsilon, H, M_{fu}$ and M_{ox}

The simplest approach to the calculation of combusting flows, and particularly diffusion flames, is to assume that chemical reactions are so fast that the process is entirely diffusion controlled. If, in addition, it is assumed that fuel and oxygen cannot appear at the same place even at different times, fluctuating properties can be neglected in the combustion model and equations solved only for the mass fractions of fuel and oxygen as well as enthalpy and aerodynamic properties. This model is known to be crude and does not justify consideration of the differences between time and mass-weighted averages and the general equation of the panel has been written in time-averaged form. Strictly speaking, this equation should contain correlations between fluctuating density and other flow properties but, again in view of the quality of the assumptions, they have been neglected.

The solution of equations for the mean mass fractions of fuel and oxygen is impossible in that the corresponding source terms would have to contain source terms to represent the infinite reaction rates. The problem is avoided by combining the two mass fractions, to form the mixture fraction f which is defined in the following panel and has been referred to on previous pages. Thus, the model requires the solution of the conservation equations for U_i, k, ε, H and f.

The value of this model has been examined, for example by Khalil *et al.* (1975) and shown to represent trends in a generally correct manner but to give quantitative results which are subject to large errors. In particular, it usually results in the overestimation of temperature in reaction regions by amounts which can exceed 500°C. It is not to be recommended for serious, quantitative calculations and its limitations may be linked directly with the second assumption of the panel and the lack of consideration of the influence of the fluctuations of scalar quantities.

15.11 Simplest combustion model, 2

- Replace M_{fu} and M_{ox} by the mixture fraction
$$f \equiv \frac{[M_{fu} - (M_{ox}/i)] - [M_{fu} - (M_{ox}/i)]_{\text{air stream}}}{[M_{fu} - (M_{ox}/i)]_{\text{fuel stream}} - [M_{fu} - (M_{ox}/i)]_{\text{air stream}}} \left[= \frac{\text{fuel–air ratio}}{1 - \text{fuel–air ratio}} \right]$$
- Determine temperature from
$$C_p T = H - M_{fu} H_{fu} + f(H_{fu} + C_p T_{\text{fuel stream}}) \\ + f(1 - f)(C_p T_{\text{fuel stream}})$$
- Specific heat represented by
$$C_{p_{\text{mix}}} = \sum_i m_i C_{p_i}$$
$$C_{p_i} = a_0 + a_1 T + a_2 T^2 + a_3 T^3 + \cdots$$

The physical and numerical difficulties associated with the source terms in the M_{fu} and M_{ox} equations may be avoided by replacing the corresponding differential equations by a single equation for f, as defined in the panel. This approach is satisfactory only for nonpremixed flames.

The panel indicates that temperatures may be determined from the definition of enthalpy, neglecting mean kinetic energy, and can only be as precise as the representation of specific heat. In general, the variation of C_p with T requires a high-order polynomial as indicated in JANAF Thermodynamic Tables (1971). The heat of reaction must also be specified correctly.

The equation for enthalpy has the same form as that for the mixture fraction f except when the source term, S_H, is finite. Thus, if S_H is zero, a solution for f is also a solution for H. In some cases, for example in furnaces other than those with very small dimensions, radiation is important and implies a finite value for S_H. Radiation is described briefly in the following paragraphs. There is considerable literature on the subject and the reader can find additional useful information, for example, in the works of Chandrasekhar (1960), Viskanta (1966), Whitacre and McCann (1975) and Truelove (1976).

The zone method of representing radiation is potentially the most precise but requires computer storage which precludes practical calculations within the framework of the conservation equations discussed here. Flux methods are, therefore, used to determine S_H and are described below.

The radiation may be represented within S_H by the equation
$$S_H = 2a(R_x + R_y - 2\sigma T^4),$$
where R_x and R_y are the positive energy fluxes in the x and y (or r) directions and a is an absorption coefficient; σ is the Stefan–Boltzmann constant. The four-flux method of Gosman and Lockwood (1972) represented R_x, R_y and a by

$$\frac{\partial}{\partial x}\frac{1}{a}\frac{\partial R_x}{\partial x} + a(\sigma T^4 - R_x) = 0,$$

$$\frac{1}{r}\frac{\partial}{\partial r}\left(\frac{r^2}{ar+1}\right)\frac{\partial R_y}{\partial r} + a(\sigma T^4 - R_y) = 0,$$

$$a = 0{\cdot}21\, M_{\text{fu}} + 0{\cdot}12\, M_{\text{pr}},$$

and a scattering term is readily included, if appropriate. This method can readily be incorporated in the enthalpy equation as shown, for example by Gosman and Lockwood (1972) and Khalil *et al.* (1975). In addition, the corresponding six flux method appropriate to three-dimensional flows has been used with considerable success, for example by Abou Ellail *et al.* (1977) who made use of the modified flux approach of Lockwood and Shah (1976).

The later coupled models of DeMarco and Lockwood (1975), Selcuk and Siddall (1976) and Lockwood and Shah (1978) are to be preferred on grounds of precision but are more difficult to incorporate in a numerical procedure. A comparison and appraisal of various radiation models has been provided by Khalil and Truelove (1978).

15.12 Improved diffusion-flame models, 1

- Retain assumptions of panel 15.10 except for "mixed is burnt"
- Assume fuel and oxygen can coexist at some location but not at same time
- Probability density distribution of f, for example

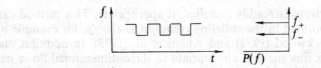

- Alternatively, a clipped Gaussian assumption may be used and temperature obtained from the equation

$$T = AT_{\text{air stream}} + BT_{\text{fuel stream}} + \frac{1}{\sigma\sqrt{2\pi}} \int_0^1 \left[\frac{H - M_{\text{fu}}H_{\text{fu}}}{C_p} \exp\left(-\frac{1}{2}\frac{f-\mu}{\sigma}\right)\right] df$$

with

$$M_{\text{fu}} = \int_0^1 M_{\text{fu}}(f) P(f) \, df$$

The assumptions that fuel and oxygen cannot coexist, panel 15.10, may be replaced by the more realistic assumption that they cannot exist at the same location at the same time but can occur at the same location at different times. This means that a time-variation of f must be determined or specified and the probability density functions of panel 15.8 allow the latter. The panel shows that the assumption of a double-delta function corresponds to a rectangular wave variation of f with time. In this case, the values of f_+ and f_- may be determined from the solution of an equation, which has the same form as that for the general variable Φ, with the rms of f-fluctuations as dependent variable. This approach was suggested by Spalding and used by Khalil et al. (1975). It is more realistic, particularly in its representation of temperature, than the model of the previous two panels, but is itself unrealistic and requires special treatment when f_+ and f_- approach 1 and 0 respectively.

The use of a normal distribution of f with time, except where it reaches 1 or 0 was proposed by Naguib and Lockwood (1976) and used by Hutchinson et al. (1977). It again involves the solution of an equation with $\overline{f'^2}$ as dependent variable and a stored function of the form

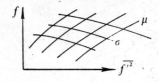

Combustion

which provide values of σ and μ corresponding to f and $\overline{f'^2}$. In this case, temperature must be evaluated from an expression of the form indicated on the panel and 1·4 times the computer time of the simple method is required.

The beta-function approach of panel 15.8 is probably to be preferred on the grounds that it is easily represented by an analytic function and as indicated by panel 15.9, for example, is in good accord with experimental information.

15.13 Improved diffusion-flame models, 2

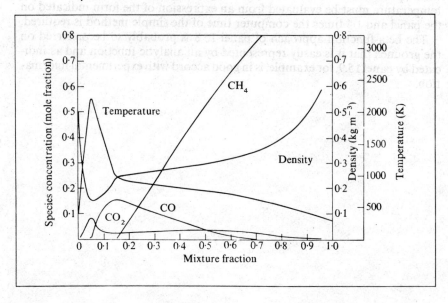

In the combustion models of the previous three slides, the heat of reaction has been considered as resulting from a single and instantaneous global reaction from fuel and oxygen to products. A preferable assumption is that of equilibrium products formed instantaneously and, for example, the equilibrium composition of a mixture of methane and air may be determined by the computer program of Gordon and McBride (1971). With this assumption, and if the source term in the enthalpy equation can be neglected, the equilibrium gas composition, temperature and density can be related to a single passive scalar, for example a mixture fraction defined as the mass of fuel, both burnt and unburnt, divided by the total mass of mixture which has also been assigned the symbol f although less restrictive than the definition of panel 15.11 which relates to a single-step reaction.

The assumptions of the previous paragraph require the solution of equations for $U_j, k, \varepsilon, \overline{f'^2}$ and, with a beta function for the probability density distribution, represent a convenient and adequate model for the calculation of a wide range of combusting flows. As with the global model approach of panel 15.12, the equations should be solved in density-weighted form and the unweighted quantities deduced subsequently, if required. This approach has been taken by, for example, Jones and McGuirk (1979b) and Atkinson (1980) and it appears that in their recirculating flows, the precision of the results is limited more by the turbulence model than the combustion model.

The above scheme still involves the assumption of instantaneous reaction which is known to be incorrect. In hydrocarbon-fuelled flames it is desirable to represent, at least, the finite rate of the reaction from CO to CO_2.

This implies at least a two-step reaction and, if the heat loss can be considered negligible, the instantaneous thermodynamic state of the gas including temperature and combustion is determinable in terms of three independent quantities. The mixture fraction f, and mass fractions of HC and CO_2 are an appropriate choice although, in practice, any three can be selected. The species formation rates are highly nonlinear functions of temperature and concentration, and the mean values of these rates have to be included in the governing transport equations before solution is possible. In the shorter term, Jones (1979) has suggested the specification of the shape of the joint probability density function for the three independent scalars in terms of all the independent first and second moments, e.g.

$$\tilde{f}, \tilde{C}_{CO_2}, \tilde{C}_{HC}, \widetilde{f''^2}, \widetilde{C''^2_{CO_2}}, \widetilde{C''^2_{HC}}, \widetilde{f''C''_{HC}}, \widetilde{f''C''_{CO_2}} \text{ and } \widetilde{C''_{HC}C''_{CO_2}}.$$

The values of these moments have then to be obtained from solution of their respective modelled transport equations. The functional form of the p.d.f. must still be known but there is indirect evidence that, provided the selected form complies with the bound restraints, the results will not be too sensitive to a specified form.

An alternative, and longer-term approach to evaluating mean species formation rates, involves the use and solution of a joint probability density function transport equation for the independent species and was referred to in connection with panel 15.8. In this formulation no approximation is required to evaluate mean formation rate though equally important terms associated with mixing appear in the equation and require approximation.

15.14 Premixed flames

- A degree of chemical control occurs in premixed flames and may be represented by an Arrhenius-type expression,
$$S_{fu} = A\rho^2 M_{ox} M_{fu} \exp(-E/RT)$$
- Physical control may also be present and depends on eddy size and time scale,
$$S_{fu} = C_R \overline{m_{fu}^2} \rho\varepsilon/k$$

In premixed and partially premixed flames, the chemistry exerts an important influence and must be considered. The source in the fuel equation may be represented by a global rate equation as indicated in the panel with the constants A and E depending upon the fuel.

Physical control may also be exerted, however, because the fuel and oxygen must mix at the microscale level to allow combustion. Thus, Spalding (1970) proposed the eddy break-up model and suggested the use of the fuel source term which resulted in the smaller source. This has been shown, for example, by Khalil et al. (1975) to work reasonably well although there is no rigorous justification for the eddy break-up model and Pope (1976), for example, has demonstrated that C_R is far from constant. It has also been used for non-premixed arrangements although it does not lead to unique solutions in this situation.

An alternative procedure, which can also be used with arbitrary fuelled flames, is due to Magnusson et al. (1978) and implies that the physically controlled reaction rate has the form
$$S_{fu} = CM_{fu}\rho\varepsilon/k(v_\varepsilon/k)^2 x$$
where
$$x = \frac{(1 - M_{fu} - M_{ox})/(1 + i)}{(1 - M_{fu} - M_{ox})/(1 + i) + M_{fu}}$$

This is simpler, in that the m_{fu}^2 equation does not have to be formulated and solved. It has not been examined in detail but shows promise, particularly for very complex systems, such as the kerosene-fuelled combustor of El Banhawy and Whitelaw (1979).

Clearly, the mixture fraction f is inappropriate to premixed flames and equations are solved for the mean mass fractions of fuel, oxygen or a reaction progress variable and, for the method of the panel, for the rms of the fluctuations of the fuel mass fraction.

The assumption of single-step reaction in a premixed flame calculation can readily be replaced by a multistep reaction and the hydrocarbon reaction scheme proposed by Waldman *et al.* (1974) is given below.

$$CH_4 + OH \leftrightarrows CH_3 + H_2O$$
$$CH_4 + H \leftrightarrows CH_3 + H_2$$
$$CH_3 + O_2 \leftrightarrows CH_2O + OH$$
$$CH_2O + O \leftrightarrows CHO + OH$$
$$CHO + OH \leftrightarrows CO + H_2O$$
$$CO + OH \leftrightarrows CO_2 + H$$
$$H + O_2 \leftrightarrows O + OH$$
$$O + H_2O \leftrightarrows OH + OH$$

Hutchinson *et al.* (1977) found that, in regions of high temperature, the use of the above reaction scheme and particularly the consideration of CO led to values of local mean temperature 40°C less than those of the single-step reaction: it may be expected that discrepancies up to around three times this figure can occur. The computer time required to solve the above kinetics equations is approximately three times that of the simplest calculation scheme and may, for many purposes, make the simpler schemes preferable.

The consideration of a multistep mechanism is likely to be important, particularly in gas-turbine applications. In such circumstances, an arbitrary-fuelled (or partially premixed) flame exists and renders the inclusion of more than a single-step reaction very difficult. One approach is discussed in relation to panel 15.15 but is far from perfect. It is likely that the use of many reactions will be impossible for many years and, in the meantime, related research will consider mainly the two-step hydrocarbon reactions with CO as an intermediate species. This statement does not apply to reactions, such as those for NO_x, which do not influence the energy-transfer processes.

15.15 Arbitrary-fuelled flames

$$S_{fu} = A\rho^2 M_{ox} M_{fu} \exp(-E/RT) + S'_{fu}$$

$$\overline{S_{fu}} = A\rho^2 M_{ox} M_{fu} \exp(-E/RT) \left[\frac{\overline{m_{ox} m_{fu}}}{M_{ox} M_{fu}} \right.$$

$$\left. + a_1 \frac{\overline{T'^2}}{T^2} + a_2 \frac{\overline{T' m_{ox}}}{T M_{ox}} + a_2 \frac{\overline{T' m_{fu}}}{T M_{fu}} + \cdots \right]$$

$$a_1 = \tfrac{1}{2}(E/RT)^2 - (E/RT); \quad a_2 = E/RT$$

Equations solved for: $H, f, M_{fu}, \overline{m_{fu} m_{ox}}, \overline{T' m_{ox}}, \overline{T' m_{fu}}$

The procedures of the previous panels are inappropriate to combustion systems which contain the features of premixed and diffusion flames. The mixture fraction f cannot be used in premixed arrangements and the eddy break-up formulation of panel 15.14 although intuitively appropriate to premixed arrangements cannot readily be justified for diffusion flames. The model of Magnusson et al. (1978), with an eddy break-up based on the mean mass fraction of fuel can be used for arbitrary-fuelled flames although it should be regarded as a first approximation. El Banhawy and Whitelaw (1979), for example, used this model in conjunction with a droplet model since the latter introduced further approximations and a simple combustion model was, therefore, appropriate.

To deal with arbitrary-fuelled flames, Borghi (1979) proposed that the Arrhenius source term should include a fluctuating component and that an expansion should be developed as shown in the panel. The truncation of the series implies that the resulting expression for the averaged source can only be valid for small values of the scalar fluctuations, for example $|T'|/T \ll 1$, and this undoubtedly limits the applicability. In addition, values for the scalar correlations must be obtained and, if the expressions are retained in time-averaged form, for the fluctuating-density correlations. This can be achieved by the solution of appropriate conservation equations in differential or algebraic form but, in either case, substantial assumptions are required. The method has been used, for example by Hutchinson et al. (1977).

15.16 Summary of some available experimental data

Reference					
Michelfelder and Lowes (1972)	6250	2000	1:11·4	Natural gas with quarl	Square section with and without swirl
Steward et al., (1972)	720	254	Co-flowing flow	Propane	Partially premixed
Baker et al., (1974)	900	300	1:5·5	Natural gas concentric jets with and without swirl	With and without swirl and combustion
Bilger and Beck (1974)	1800	305	Co-flowing flow	Hydrogen	
Beltagui and MacCullum (1975)	1400 900	460 225	1:5 1:2·5	Town gas Single jet	Isothermal and premixed flame
Bowman and Cohen (1975)	1210	122·3	1:1·3	Methane concentric jets	
Cernansky and Sawyer (1975)	356	58	Co-flowing flow	Propane	Swirl in air stream
Lenze et al., (1975)	2500	450	Co-flowing flow	Natural gas Town gas	
Wu and Fricker (1976)	5000	900	1:6·87	Natural gas concentric jets with quarl	With and without swirl and combustion

The table provides a summary of some information which has and can be used for the evaluation and improvement of combustion models. Several more recent furnaces are presently in use and more and improved information can confidently be expected. Information is also available on liquid-spray flames and some has been referred to, for example, by El Banhawy and Whitelaw (1979). There is almost no detailed information of solid-fuelled flames.

It should be emphasized that measurements of the properties of combusting flows are subject to uncertainties which are usually larger than those in isothermal flows. In addition, measurements in oil and solid fuelled flames are less exact than those in gaseous flames. Discussions of measurement accuracy can be found, for example, in the volume edited by Kennedy (1977) and it should be appreciated that the evaluation of calculation methods is limited by uncertainties which can be of the order of ± 2 per cent in U_i, ± 5 per cent in T, ± 5 per cent in the mass fractions of major species and larger amounts for correlations.

References

ABOU ELLAIL, M. M. M., GOSMAN, A. D., LOCKWOOD, F. C. and MEGAHED, I.E.A. (1977). A three dimensional procedure for combustion chamber flows. *AIAA Paper*, 77–138.

AHMAD, Q. A., LUXTON, R. E. and ANTONIA, R. A. (1976). Characteristics of a turbulent boundary layer with an external turbulent uniform shear flow. *J. Fluid Mech.* **77**, 369.

ALBER, I. E. and LEES, L. (1968). Integral theory for supersonic turbulent base flows. *AIAA J.* **6**, 1343.

ANDREOPOULOS, J. and BRADSHAW, P. (1980). Measurements of interacting turbulent shear layers in the near wake of an aerofoil. *J. Fluid Mech.* **100**, 639.

ARIELI, R. and MURPHY, J. D. (1980). Pseudo-direct solutions to the boundary-layer equations for separated flows, *AIAA J.*, **18**, 883.

ATKINSON, K. N. (1980). Predictions of local flow properties in an axisymmetric disc-stabilised combustor. Imperial College, Mech. Engng. Dept. Report FS/80.

BAKER, R. J., HUTCHINSON, P., KHALIL, E. E. and WHITELAW, J. H. (1974). Measurements of three velocity components in a model furnace with and without combustion. 15th Symposium (International) on Combustion Proc., p. 553. The Combustion Institute, U.S.A.

BALSA, T. F. and MELLOR, G. L. (1975). The simulation of axial compressor performance using an annulus wall boundary-layer theory. *J. Engng. Power*, **97A**, 305.

BECKER, H. A., HOTTEL, H. C. and WILLIAMS, G. G. (1967). On the light scatter technique for the study of turbulence and mixing. *J. Fluid Mech.* **30**, 259.

BELCHER, R. J., BURGGRAF, O. R., COOKE, J. C., ROBINS, A. J. and STEWARTSON, K. (1971). Limitless boundary layers. *In* "Recent Research on Unsteady Boundary Layers". (Eichelbrenner, E. A., ed.), p. 1444. Laval University Press, Quebec.

BELTAGUI, S. A. and MACCALLUM, N. R. L. (1975). The aerodynamics and modelling of vane-swirled premixed flames in furnaces. International Flame Research Document No. F21/CA/14.

BERGELES, G., GOSMAN, A. D. and LAUNDER, B. E. (1978). The prediction of three-dimensional discrete-hole cooling process. *Num. Heat Transf.*, **1**, 217.

BERGER, S. A. (1971). Laminar Wakes. American Elsevier, New York.

BILGER, R. W. (1977). Probe measurements in turbulent combustion. *Prog. Astronaut Aeronaut* **53**, 49.

BILGER, R. W. and BECK, R. F. (1974). Further experiments on turbulent jet diffusion flames. University of Sydney, Mech. Engng. Dept. Report No. TN F-67.

BISSONNETTE, L. R. and MELLOR, G. L. (1974). Experiments on the behavior of an axisymmetric turbulent boundary layer with a sudden circumferential strain. *J. Fluid Mech.* **63**, 369.

BORGHI, R. (1979). Models of turbulent combustion for numerical predictions. ONERA Report T. P. 1979–1.

BOWER, W. W. and PETERS, G. R. (1979). A Navier–Stokes scheme for the calculation of three-dimensional impinging jet flows. Proc. 2nd International Symposium Turbulent Shear Flows, p. 14. 17, Imperial College, London.

BOWMAN, C. T. and COHEN, L. S. (1975). Influence of aerodynamic phenomena on pollutant formation in combustion. Vol. 1, "Experimental Results". EPA-650/2-75-061-a, USA.

BRADSHAW, P. (1970). Prediction of the turbulent near wake of a symmetrical aerofoil. *AIAA J.* **8**, 1507.

BRADSHAW, P. (1973). Effect of streamline curvature on turbulent flows. AGAR-Dograph 169.

BRADSHAW, P. (1975a). "An Introduction to Turbulence and Its Measurement". Pergamon Press, Oxford.

BRADSHAW, P. (1975b). Review—complex turbulent flows. *J. Fluids Engng.*, **971**, 146.

BRADSHAW, P. (ed.) (1976). Topics in applied physics. Vol. 12—"Turbulence". Springer, Heidelberg.

BRADSHAW, P. (1977a). Compressible turbulent shear layers. *A. Rev. Fluid Mech.* **9**, 33.

BRADSHAW, P. (1977b). Effect of external disturbances on the spreading rate of a plane, turbulent jet. *J. Fluid Mech.* **80**, 795.

BRADSHAW, P. and UNSWORTH, K. (1977). Computation of complex turbulent flows. In "Reviews in Viscous Flows" Lockheed-Georgia Co. Report LG77ER0044.

BRADSHAW, P., DEAN, R. B. and MCELIGOT, D. M. (1973). Calculation of interacting turbulent shear layers-duct flow. *J. Fluids Engng.* **951**, 214.

BRADSHAW, P., FERRISS, D. H. and ATWELL, N. P. (1967). Calculation of boundary-layer development using the turbulent energy equation. *J. Fluid Mech.* **28**, 593.

BRILEY, W. R. (1971). A numerical study of laminar separation bubbles using the Navier–Stokes equations. *J. Fluid Mech.* **47**, 713.

BRILEY, W. R. and MCDONALD, H. (1973). An implicit numerical method for the multi-dimensional compressible Navier–Stokes equations. United Aircraft Research Laboratories Report M911363–6.

BRILEY, W. R. and MCDONALD, H. (1974). Numerical prediction of incompressible separation bubbles. *J. Fluid Mech.* **69**, 631.

BRILEY, W. R., MCDONALD, H. and GIBELING, H. J. (1977). Solution of the multi-dimensional compressible Navier–Stokes equations by a generalised implicit method. *J. comp. Phys.* **24**, 372.

BROWN, S. N. (1965). Singularities associated with separating boundary layers. *Phil. Trans. R. Soc., London*, A257, 409.

BROWN, W. B. (1961). A stability criterion for three-dimensional boundary layers. In "Boundary Layer and Flow Control" (Lachmann, ed. G. V.), Pergamon p.904.

BRYANT, D. and HUMPHREY, J. A. C. (1976). Conservation equations for laminar and turbulent flows in general three-dimensional curvilinear coordinates. Imperial College, Mech. Engng. Dept. Report CHT/76/6.

BUSHNELL, D. M., CARY, A. M. and HOLLEY, B. B. (1975). Mixing length in low Reynolds number compressible turbulent boundary layers. *AIAA J.* **13**, 1119.

CARMODY, T. (1964). Establishment of the wake behind a disc. *J. basic Engng.* **86**, 869.

CARTER, J. E. (1974). Solutions for laminar boundary layers with separation and reattachment. *AIAA* Paper No. 74-583.

CARTER, J. E. (1975). Inverse solutions for laminar boundary-layer flows with separation and reattachment. NASA TR R-447.

CARTER, J. E. and WORNOM, S. F. (1975). A forward marching procedure for separated boundary-layer flows. *AIAA J.* **13**, 8, 1101.

CASTRO, I. (1973). A highly distorted turbulent free-stream layer. Ph.D. Thesis, Imperial College, London University.

CASTRO, I. P. (1979). Numerical difficulties in the calculation of complex turbulent flows. *In* "Turbulent Shear Flows", **1**, 220. Springer Verlag, Berlin.

CASTRO, I. P. and BRADSHAW, P. (1976). The turbulence structure of a highly curved mixing layer. *J. Fluid Mech.* **73**, 265.

CATHERALL, D. and MANGLER, K. W. (1966). The integration of the two-dimensional laminar boundary-layer equations past the point of vanishing skin friction. *J. Fluid Mech.* **26**, 163.

CAUGHEY, D. A. and JAMESON, A. (1979). Numerical calculation of transonic potential flow about wing-body combinations *AIAA J.* **17**, 175.

CEBECI, T. (1974). Calculation of three-dimensional boundary layers. I. Swept infinite cylinders and small cross-flow. *AIAA J.* **12**, 779.

CEBECI, T. (1975). Calculation of three-dimensional boundary layers. II. Three-dimensional flows in Cartesian coordinates. *AIAA J.* **13**, 1056.

CEBECI, T. (1976a). An inverse boundary-layer method for compressible laminar and turbulent flows. *J. Aircraft,* **13**, 709.

CEBECI, T. (1976b). Separated flows and their representation by boundary-layer equations. Mech. Engng Dept. Report No. TR-76-1, California State University, Long Beach.

CEBECI, T. (1978). An unsteady laminar boundary layer with separation and reattachment. *AIAA.* **16**, 1305.

CEBECI, T. (1979). The laminar boundary layer on a circular cylinder started impulsively from rest. *J. comp. Phys.* **31**, 153.

CEBECI, T. and ABBOTT, D. E. (1975). Boundary layers on a rotating disk. *AIAA J.* **13**, 829.

CEBECI, T. and BRADSHAW, P. (1977). "Momentum Transfer in Boundary Layers". Hemisphere-McGraw Hill, Washington.

CEBECI, T. and CARR, W. L. (1978). A computer program for calculating laminar and turbulent boundary layes for two-dimensional time dependent flows. NASA TM 78470.

CEBECI, T. and CARR, W. L. (1980). Computation of unsteady boundary layers with flow reversal and evaluation of two separate turbulence models. In preparation.

CEBECI, T. and CHANG, K. C. (1978). A general method for calculating momentum and heat transfer in laminar and turbulent duct flows. *Num. Heat Trans.* **1**, 39.

CEBECI, T. and KELLER, H. B. (1972). Laminar boundary layers with assigned wall shear. Lecture Notes in Physics. (Proc. Int. Cong. Numer. Methods Fluid Dynamics, 3rd), Vol. 19. Springer-Verlag, Berlin and New York.

CEBECI, T. and KELLER, H. B. (1974). Flows in ducts by boundary-layer theory.

Proceedings of Fifth Australasian Conf. on Hydraulics and Fluid Mech. University of Canterbury, Christchurch, New Zealand.

CEBECI, T. and KELLER, H. B. (1977). Stability calculations for a rotating disk. AGARD CP 224, Paper 7.

CEBECI, T. and MEIER, H. U. (1979). Modelling requirements for the calculation the turbulent flow around airfoils, wings, and bodies of revolution. Turbulent Boundary Layers-Experiments, Theory and Modelling, AGARD CP 271, Paper 16.

CEBECI, T. and SMITH, A. M. O. (1974). "Analysis of Turbulent Boundary Layers". Applied Mathematics and Mechanics, Vol. 15. Academic Press. New York.

CEBECI, T. and STEWARTSON, K. (1978). Unpublished work.

CEBECI, T. and STEWARTSON, K. (1980a). On stability and transition in three-dimensional flows. *AIAA J.* **18**, 398.

CEBECI, T. and STEWARTSON, K. (1980b). On the prediction of transition in three--dimensional flows. Laminar/Turbulent Transition, 243, Springer Verlag.

CEBECI, T., CARR, W. L. and BRADSHAW, P. (1979). Prediction of unsteady turbulent boundary layers with flow reversal. Proc. 2nd Symposium on Turbulent Shear Flows, p. 14.23. Imperial College, London.

CEBECI, T., CHANG, K. C. and BRADSHAW, P. (1980). Solution of a hyperbolic system of turbulence-model equations by the "Box" method. *Comp. Meth. Appl. Mech. Engng*, **22**, 213.

CEBECI, T., CHANG, K. C. and KAUPS, K. (1980). A general method for calculating three-dimensional laminar and turbulent boundary layers on ship hulls. Ocean Engineering, **7**, 229.

CEBECI, T., KAUPS, K. and MOSER, A. (1976). Calculation of three-dimensional boundary layers. III. Three-dimensional incompressible flows in curvilinear orthogonal coordinates. *AIAA J.* **14**, 1090.

CEBECI, T., KAUPS, K. and RAMSEY, J. A. (1977). A general method for calculating three-dimensional compressible laminar and turbulent boundary layers on arbitrary wings. NASA CR 2777.

CEBECI, T., KELLER, H. B. and WILLIAMS, P. G. (1979). Separating boundary-layer calculations. *J. comp. Phys.* **31**, 363.

CEBECI, T., KHALIL, E. E. and WHITELAW, J. H. (1979). The calculation of separated boundary layers. *AIAA J.* **17**, 1291.

CEBECI, T., KHATTAB, A. A. and STEWARTSON, K. (1979). Prediction of three-dimensional laminar and turbulent boundary layers on bodies of revolution at high angles of attack. Proc. 2nd International Symposium on Turbulent Shear Flows, p. 15.8. Imperial College, London.

CEBECI, T., KHATTAB, A. A. and STEWARTSON, K. (1980a). On Nose Separation. *J. Fluid Mech.* **97**, 435.

CEBECI, T., KHATTAB, A. A. and STEWARTSON, K̃. (1980b). Three-dimensional laminar boundary layers and the Ok of accessibility. To be published in *J. Fluid Mech.*

CEBECI, T., MOSINSKIS, G. J. and SMITH, A. M. O. (1972a). Calculation of separation points in turbulent flows. *J. Aircraft*, **9**, 618.

CEBECI, T., MOSINSKIS, G. J. and SMITH, A. M. O. (1972b). Calculation of viscous drag of two-dimensional and axisymmetric bodies in incompressible flows. *J. Aircraft*, **9**, 691.

CEBECI, T., SMITH, A. M. O. and MOSINSKIS, G. (1970). Solution of the incompressible turbulent boundary-layer equations with heat transfer. *J. Heat Transf.* **92C**, 133.

CEBECI, T., STEWARTSON, K. and WILLIAMS, P. G. (1979). On the response of a stagnation boundary layer to a change in the external velocity. *SIAM J. Appl. Math.* **36**, 190.

CEBECI, T., KAUPS, K., MOSINSKIS, G. J. and REHN, J. A. (1973). Some problems of the calculation of three-dimensional boundary-layer flows on general configurations. NASA Report CR-2285.

CEBECI, T., THIELE, F., WILLIAMS, P. G. and STEWARTSON, K. (1979). On the calculation of symmetric wakes. I. Two-dimensional flows. *Num. Heat Transf.* **2**, 35.

CERNANSKY, N. P. and SAWYER, R. F. (1975). NO and NO_2 formation in turbulent hydrocarbon/air diffusion flame. Proc. 15th International Symposium on Combustion, Combustion Institute, U.S.A., p. 1039.

CHAM, T. S. and HEAD, M. R. (1969). Turbulent boundary-layer flows on a rotating disk. *J. Fluid Mech.* **37**, 129.

CHANDRASEKHAR, S. (1960). "Radiation Transfer". Dover Publications, New York.

CHEVRAY, R. and KOVASZNAY, L. S. G. (1969). Turbulent measurements in the wake of a thin flat plate. *AIAA J.* **7**, 1641.

CHOE, H., KAYS, W. M. and MOFFAT, R. J. (1975). The turbulent boundary layer on a full-coverage film-cooled surface: An experimental heat transfer study with normal injection. NACA CR-2642.

COLES, D. (1956). The law of the wake in the turbulent boundary layer. *J. Fluid Mech.* **1**, 191.

COLES, D. and HIRST, E. A. (ed.) (1969). Proceedings, Computation of Turbulent Boundary Layers—1968 AFOSR-IFP—Stanford Conference, Vol 2, Thermosciences Division, Stanford University.

COMTE-BELLOT, G., and CORRSIN, S. (1966). The use of a contraction to improve the isotropy of grid turbulence. *J. Fluid Mech.* **25**, 657.

CRABB, D. (1979). Jets in crossflow. Ph.D. Thesis, Imperial College, London University.

CRANK, J. and NICHOLSON, P. (1947). A practical method for numerical evaluation of solutions of partial-differential equations of the heat-conduction type *Proc. Camb. Phil. Soc.*, **43**, 50.

CRAWFORD, M. D., KAYS, W. M. and MOFFAT, R. J. (1976). Heat transfer to a full-coverage film-cooled surface with 30-deg slant-hole injection. NASA CR-2786.

CROW, S. C. (1968). Viscoelastic properties of fine-grained incompressible turbulence. *J. Fluid Mech.* **33**, 1.

DALY, B. J. and HARLOW, F. H. (1970). Transport equations in turbulence. *Phys. Fluids*, **13**, 2634.

DAVIDOV, B. I. (1961). On the statistical dynamics of an incompressible turbulence fluid. *Dokl. Akad Nauk SSSR*, **136**, 47.

DEAN, R. B. (1974). An investigation of shear-layer interaction in ducts and diffusers. Ph.D. Thesis, Imperial College, London University; see also *J. Fluid Mech.* **78**, 642 (1976).

DEAN, R. C. (1974). The fluid dynamic design of advanced centrifugal compressors. TN-185, Creare, Inc., Hanover, New Hampshire.

DEMARCO, A. M. and LOCKWOOD, F. C. (1975). A new flux model for the calculation of three-dimensional radiation heat transfer. La Revista dei Combustible, **29**, 184.

DONALDSON, C. du P. (1969). A computer study of boundary-layer transition. *AIAA J.* **7**, 271.

DOUGLAS, J., JR. (1955). On the numerical integration of $\partial^2 u/\partial x^2 + \partial^2 u/\partial y^2 = \partial u/\partial t$ by implicit methods. *J. Soc. Ind. Appl. Math.* **3**, 42.

DRING, R. P. (1971). A momentum-integral analysis of the three-dimensional turbine end-wall boundary layer. *J. Engng Power*, **93A**, 386.

DURÃO, D. F. G. and WHITELAW, J. H. (1974). Measurements in the region of recirculation behind a disc. Imperial College, Mech. Engng. Dept., Report HTS/74/14, also Proc. 2nd Laser Velocimeter Workshop, Purdue University, **2**, 413.

DURÃO, D. F. G. and WHITELAW, J. H. (1977). Velocity characteristics of the flow in the near-wake of a disc. *J. Fluid Mech.* **85**, 369.

EAST, L. F. and SAWYER, W. G. (1979). An investigation of the structure of equilibrium turbulent boundary layers Turbulent Boundary Layers—Experiments, Theory, Modelling. AGARD CP 271.

ECKARDT, D. (1975). Instantaneous measurements in the jet-wake discharge flow of a centrifugal compressor impeller. *J. Engng Power*, **97**, 347; also *J. Fluids Engng*, **981**, 390.

EL BANHAWY, Y. and WHITELAW, J. H. (1979). Assessment of an approach to the calculation of the flow properties of spray flames. AGARD CP 275, Paper 12.

EISEMAN, P. R., LEVY, R., MCDONALD, H. and BRILEY, W. R. (1978). Development of a three-dimensional turbulent duct flow analysis. NASA CR 3029.

FAVRE, A. (1965). Equations des gaz turbulents combustibles. *J. Phys. Atmos.* **4**, 361.

FERZIGER, J. H. (1977). Large eddy numerical simulations of turbulent flow. *AIAA J.* **15**, 1261.

FERNHOLZ, H. H. and FINLEY, P. J. (1977). A critical compilation of compressible turbulent boundary layer data. AGARDograph 223.

GALBRAITH, R. A. and HEAD, M. R. (1975). Eddy viscosity and mixing length from measured boundary-layer developments. *Aeronaut. Q.* **26**, 133.

GARABEDIAN, P. (1964). "Partial Differential Equations". Interscience Publishers, New York.

GENTRY, R. A., MARTIN, R. E. and DALY, B. J. (1966). An Eulerian differencing method for unsteady compressible flow problems. *J. comp. Phys.* **1**, 87.

GASTER, M. (1975). A theoretical model of a wave packet in a boundary layer on a flat plate. *Proc. R. Soc., London, A347*, 271.

GESSNER, F. B. and EMERY, A. F. Numerical prediction of developing turbulent flow in rectangular ducts. Proc. 2nd International Symposium on Turbulent Shear Flows, p.17.1, Imperial College, London.

GIBSON, M. M. (1978). An algebraic stress and heat-flux model for turbulent shear flow with streamline curvature. *Int. J. Heat Mass Transf.* **21**, 1609.

GIBSON, M. M. (1979). Prediction of curved free shear layers with a Reynolds-stress model of turbulence. Proc. 2nd International Symposium on Turbulent Shear Flow, 2.6, Imperial College, London.

GIBSON, M. M. and LAUNDER, B. E. (1976). On the calculation of horizontal turbulent shear flows under gravitational influence. *J. Heat Transf.* **98**, 81.

GOLDSCHMIDT, V. and BRADSHAW, P. (1973). Flapping of a plane jet. *Phys. Fluids*, **16**, 354.

GORDON, S. and MCBRIDE, B. J. (1971). Computer program for calculation of complex equilibrium composition, rocket performance, incident and reflected shocks and Chapman–Jouguet deteriations. NASA SP-273.

GOSMAN, A. D. and LOCKWOOD, F. C. (1972). Incorporation of a flux model for radiation into a finite/difference procedure for furnace calculations. Proc. 14th Symposium (International) on Combustion, 661.

GOSMAN, A. D. and PUN, W. M. (1973). Calculation of recirculating flows. Imperial College, Mech. Engng Dept. Report HTS/73/2.

GOSMAN, A. D. and WATKINS, A. P. (1978). Predictions of local instantaneous heat transfer in motored engines. Imperial College, Mech. Engng Dept. Report FS/78/37.

GOSMAN, A. D., KHALIL, E. E. and WHITELAW, J. H. (1979). The calculation of two-dimensional turbulent recirculating flows. *In* "Turbulent Shear Flows" Springer-Verlag, Berlin, **1**, 287.

GOSMAN, A. D., LOCKWOOD, F. C. and SALUJA, A. P. (1978). The prediction of cylindrical furnaces gaseous fuelled with premixed and diffusion burners. Proc. 17th Symposium (International) on Combustion, 747.

GOSMAN, A. D., PUN, W. M., RUNCHAL, A. K., SPALDING, D. B. and WOLFSHTEIN, M. (1969). "Heat and Mass Transfer in Recirculating Flows". Academic Press, London and New York.

GREEN, J. E. (1970). Interaction between shock waves and turbulent boundary layers. *Prog. Aeronaut. Sci.* **11**, 235.

HABIB, M. A. and WHITELAW, J. H. (1979). Velocity characteristics of a confined coaxial jet. *J. Fluids Engng.* **101**, 521.

HABIB, M. A. and WHITELAW, J. H. (1980). Measured velocity characteristics of confined coaxial jets with and without swirl. *J. Fluids. Engng.*, **102**, 47.

HANJALIC, K. and LAUNDER, B. E. (1972). A Reynolds stress model of turbulence and its application to thin shear flows. *J. Fluid Mech.* **52**, 609.

HANJALIC, K. and LAUNDER, B. E. (1976). Contribution towards a Reynolds stress closure for low-Reynolds number turbulence. *J. Fluid Mech.* **74**, 593.

HANJALIC, K. and LAUNDER, B. E. (1979). Preferential spectral transport by irrotational straining. Proc of 2nd International Symposium on Tubulent Shear Flow, Imperial College, London.

HARLOW, F. H. and AMSDEN, A. A. (1971). A numerical fluid dynamics calculation method for all flow speeds. *J. comp. Phys.* **8**, 197.

HARLOW, F. M. and NAKAYAMA, P. I. (1967). Turbulence transport equations. *Physics Fluids*, **10**, 2323.

HEAD, M. R. (1958). Entrainment in the turbulent boundary layer. Aeronaut. Res. Council R & M 3152.

HERZIG, H. Z., HANSEN, A. G. and COSTELLO, G. R. (1953). A visualization study of secondary flows in cascades. NACA Report 1163.

HESS, J. L. (1979). A higher order panel method for three-dimensional potential flow. McDonnell Douglas Report MDC J8519.

HESS, J. L. and SMITH, A. M. O. (1967). Calculations of potential flow about arbitrary bodies. *Prog. Aeronaut. Sci.* **8**, 1.

HIGUCHI, H. and RUBESIN, M. W. (1978). Behaviour of a tubulent boundary layer subjected to sudden transverse strains. *AIAA* paper 78–201.

HIRSH, R. S. and CEBECI, T. (1977). Calculation of three-dimensional boundary layers with negative-crossflow on bodies of revolution. *AIAA* paper 77–683.

HORLOCK, J. H. and PERKINS, H. J. (1974). Annulus wall boundary layers in turbomachines. AGARDograph 185.

HORTON, H. P. (1974). Separating laminar boundary layers with prescribed wall shear. *AIAA J.* **12**, 1772.

HOWARD, J. H. G., PRATAP, V. S. and SPALDING, D. B. (1975). Measurements of the turbulent flow in a rectangular curved duct. Imperial College, Mech. Engng. Dept. Report HTS/75/18.

HUANG, T. T., SANTELLI, N. and BELT, G. (1978). Stern boundary-layer flow on axisymmetric bodies. Proc. 12th Symposium on Naval Hydrodynamics, p. 127. Washington, D.C.

HUFFMAN, G. D. and BRADSHAW, P. (1972). A note on von Karman's constant in low Reynolds number turbulent flows. *J. Fluid Mech.* **53**, 45.

HUMPHREY, J. A. C., TAYLOR, A. M. K. and WHITELAW, J. H. (1977). Laminar flow in a square duct of strong curvature. *J. Fluid Mech.* **83**, 509.

HUMPHREY, J. A. C. and WHITELAW, J. H. (1977). Measurements in curved flows. *In* "Turbulence in Internal Flows" (Murthy, S. N. B., Ed.) Hemisphere, Washington.

HUMPHREY, J. A. C. (1978). Numerical calculations of developing laminar flow in pipes of arbitrary curvature radius. *Can. J. chem. Engng.*, **56**, 151.

HUMPHREY, J. A. C., WHITELAW, J. H. and YEE, G. (1980). Turbulent flow in a square duct with strong curvature. To be published in *J. Fluid Mech.*

HUTCHINSON, P., KHALIL, E. E. and WHITELAW, J. H. (1977). The measurement and calculation of furnace-flow properties. *AIAA J. Energy*, **1**, 212.

INMAN, P. N. and BRADSHAW, P. (1980). Mixing length in low Reynolds number turbulent boundary layers. Submitted to *AIAA J.*

IRWIN, H. P. A. H. (1973). Measurements of a self-preserving plane wall jet in a positive pressure gradient. *J. Fluid Mech.* **61**, 33.

IRWIN, H. P. A. H. (1974). Measurements in blown boundary layers and their prediction by Reynolds stress modelling. Ph.D. Thesis, McGill University.

ISAACSON, E. and KELLER, H. B. (1966). "Analysis of Numerical Methods". John Wiley, New York.

JAFFE, N. A., OKAMURA, T. T. and SMITH, A. M. O. (1970). Determination of spatial amplification factors and their application to predicting transition. *AIAA J.* **8**, 301.

JANAF thermochemical tables (1971). National Bureau of Standards, USA NSRDS 37, COM-71-50363.

JONES, W. P. (1971). Laminarization in strongly accelerated boundary layers. Ph.D. Thesis, Imperial College, London University.

JONES, W. P. (1979). Production methods for turbulent flows. Von Karman Institute for Fluid Dynamics, Lecture series 1979–2. *See also* Models for turbulent flows with variable density. *In* "Prediction Methods for Turbulent Flows" (W. Kollmann, ed.) (1980). Hemisphere Publishing Corporation, Washington, D.C.

JONES, W. P. and LAUNDER, B. E. (1972). The prediction of laminarization with a two-equation model of turbulence. *Int. J. Heat Mass Transf.* **15**, 301.

JONES, W. P. and MCGUIRK, J. (1979a). Computation of a round turbulent jet discharging into a confined cross flow. Proc. 2nd International Symposium on Turbulent Shear Flow, Imperial College, London.

JONES, W. P. and MCGUIRK, J. (1979b). Mathematical modelling of gas turbine combustion chambers. AGARD CP 275, Paper 4.

JONES, W. P. and PRIDDIN, C. H. (1978). Prediction of the flow field and local gas combustion in gas turbine combustors. Proc. 17th International Combustion Symposium, 399.

JONES, W. P. and RENZ, U. (1974). Condensation from a turbulent stream onto a vertical surface. *Int. J. Heat Mass Transf.* **17**, 1019.

JONES, W. P. and WHITELAW, J. H. (1978). Coupling of turbulence and chemical reaction. Proc. Workshop on Modelling of Combustion in Practical Systems, Los Angeles.

KACKER, S. C. and WHITELAW, J. H. (1968). Some properties of the two-dimensional turbulent wall jet in a moving stream. *J. Appl. Mech.* **35**, 641.

KADER, B.A. and YAGLOM, A. M. (1977). Turbulent heat and mass transfer from a wall with parallel roughness ridges. *Int. J. Heat Mass Transf.* **2**, 345.

KAPLAN, R. E. (1964). The stability of laminar incompressible boundary layers in the presence of compliant boundaries. MIT Rept ASRL-TR-116-1.

KELLER, H. B. (1968). "Numerical Methods for Two-Point Boundary-Value Problems". Ginn-Blaisdell, Waltham, Massachusetts.

KELLER, H. B. (1970). A new difference scheme for parabolic problems. In "Numerical Solution of Partial-Differential Equations". Bramble, J. (ed.), vol. II, Academic Press, New York.

KELLER, H. B. (1974). Accurate difference methods for two-point boundary-value problems. *SIAM J. numerical Analysis*, **11**, 305.

KELLER, H. B. and CEBECI, T. (1972). An inverse problem in boundary-layer flows: numerical determination of pressure gradient for a given wall shear. *J. comp. Phys.* **10**, 151.

KENT, J. H. and BILGER, R. W. (1972). Measurements in turbulent jet diffusion flames. University of Sydney Report F-41.

KENNEDY, L. A. (ed.) (1977) Turbulent Combustion. *Prog. Astronaut. Aeronaut.* **58**.

KHALIL, E. E. and TRUELOVE, J. S. (1978). Calculation of radiative heat transfer in a large gas-fired furnace. *Lett. J. Heat Mass Transf.* **4**, 353. See also AERE R 8747, 1977.

KHALIL, E. E., SPALDING, D. B. and WHITELAW, J. H. (1975). The calculation of local flow properties in two-dimensional furnaces. *Int. J. Heat Mass Transf.* **18**, 775.

KLEBANOFF, P. S. (1955). Characteristics of turbulence in a boundary layer with zero pressure gradient. NACA Rept. 1247.

KLINEBERG, J. M., and STEGER, J. L. (1974). On laminar boundary-layer separation. *AIAA* Paper No. 74–94.

KOLBE, C. D. and BOLTZ, F. W. (1951). The forces and pressure distribution at subsonic speeds on a plane wing having 45° of sweepback, and aspect ratio of 3, and a taper ratio of 0.5. NACA RM A51G31.

KOLMOGOROV, A. N. (1942). Equations of turbulent motion in an incompressible fluid. *Izv. Akad. Nauk SSSR, Ser. Fiz.* 6, 56 (Translated as Imperial College Mech. Eng. Dept. Report ON/6, 1968.)

KRAUSE, E., HIRSCHEL, E. H. and BOTHMANN, Th. (1969). Die numerische Integration der bewegungsgleichungen dreidimensionaler laminarer kompressibler Grenzschichten. Fachtagung Aerodynamik, Berlin, 1968, DGLR-Fachzeitschrift, Band 3.

LAKSHMINARAYANA, B., BRITSCH, W. R. and GEARHART, W. S. (ed.) (1974). Fluid mechanics, acoustics and design of turbomachinery. NASA SP-304.

LANGSTON, L. S., NICE, M. L. and HOOPER, R. M. (1977). Three-dimensional flow within a turbine cascade passage. *J. Engng. Power*, **99**, 218.

LAUNDER, B. E. (1976). Heat and mass transport by turbulence. Chapter 6 *in* "Topics in Applied Physics", Vol. 12—Turbulence (Bradshaw, P. ed.), Springer, Berlin.

LAUNDER, B. E. and MORSE, A. (1979). Numerical prediction of axisymmetric free shear flows with a Reynolds stress closure proceedings. *In* "Turbulent Shear Flows", **1**, 279. Springer-Verlag, Berlin.

LAUNDER, B. E. and SHARMA, B. I. (1974). Application of the energy-dissipation model of turbulence to flow near a spinning disc. *Lett. Heat Mass Transf.* **1**, 131.

✓LAUNDER, B. E. and SPALDING, D. B. (1972). "Mathematical Models of Turbulence". Academic Press, London and New York.

LAUNDER, B. E. and SPALDING, D. B. (1974). The numerical computation of turbulent flow. *Comp. Meth. Appl. Mech. Engng*, **3**, 269.

LAUNDER, B. E. and YING, W. M. (1973). The prediction of flow and heat transfer in ducts of square cross section. *Proc. Inst. Mech. Eng.* **187**, 37.

LAUNDER, B. E., MORSE, A., SPALDING, D. B. and RODI, W. (1973). The prediction of free shear stress flows–a comparison of the performance of six turbulence models. Proc. Langley Free Shear Flows Conference, NASA SP321.

LAUNDER, B. E., PRIDDIN, C. H. and SHARMA, B. I. (1977). The calculation of turbulent boundary layers on curved and spinning surfaces. *J. Fluids Engng*, **99**, 237.

LAUNDER, B. E., REECE, G. J. and RODI, W. (1975). Progress in the development of a Reynolds stress turbulence closure. *J. Fluid Mech.* **68**, 537.

LAWACZECK, O., BUTEFISH, K. A. and HEINEMANN, H. J. (1976). Vortex streets in the wakes of subsonic and transonic turbine cascades. *Revue Fr. Méc.* Supplement, p. 9 (also IAA abstract item A77-20154).

LENZE, B., MILANO, M. E. and GUNTHER, R. (1975). The mutual influence of multiple jet diffusion flame. 3rd Members Conference of IFRF (see also *Comb. Sci. Tech.*, **11**, 1).

LEONARD, B. P. (1979). A survey of finite differences of opinion on numerical muddling of the incomprehensible defective confusion equation. ASME, AMD Vol. 34, Finite Element Methods for Convection Dominated Flows.

LIEBECK, R. H. (1978). Design of subsonic airfoils for high lift. *J. Aircraft*, **15**, 547.

LOCKWOOD, F. C. and SHAH, N. G. (1976). An improved model for the calculation of radiation heat transfer in combustion chambers. ASME Paper 76-HT-55.

LOCKWOOD, F. C. and SHAH, N. G. (1978). Evaluation of an efficient flux model for furnace prediction procedures. Proc. 6th International Heat Transfer Conference, Paper EC6.

LOHMANN, R. P. (1976). The response of a developed turbulent boundary layer to local transverse surface motion. *J. Fluids Engng.*, **98**, 354.

LUMLEY, J. L. (1972). A model for computation of stratified turbulent flows. Proc. International Symposium on Stratified Flow, Novosibirsk.

LUMLEY, J. L. and KHAJEH NOURI, B. J. (1974). Computational modelling of turbulent transport. *Adv. Geophys.* **18A**, 169.

LUMLEY, J. L. (1979). Second order modelling of turbulent flows von Karman Institute for Fluid Dynamics. Lecture Series 1979–2.

LYNCH, F. T. (1978). Recent applications of advanced computational methods in the aerodynamic design of transport aircraft configurations. *Aeronaut. J. (J. R. Aeronaut. Soc.)*, **82**, 503.

MACK, L. M. (1977). Transition prediction and linear stability theory. Laminar-Turbulent Transition. AGARD CP 224.

MAGNUSSON, B. F., HJERTAGER, B. M., OLSEN, J. G. and BHADURI, D. (1978). Effects of turbulent structure and local concentrations on root formation and combustion in C_2H_2 diffusion flames. Proc. 17th Combustion Symposium.

MAHGOUB, H. and BRADSHAW, P. (1979). Calculations of turbulent-inviscid flow interactions with large normal pressure gradients. *AIAA J.* **17**, 1025.

MAJUMDAR, A. K. and SPALDING, D. B. (1977). A numerical investigation of three-dimensional flows in a rotating duct by a partially-parabolic procedure. Imperial College Report HTS/77/26 (also ARC 37189).

MATTHEWS, L. and WHITELAW, J. H. (1972). Some characteristics of the flow downstream of a blunt trailing edge in the presence of a neighboring wall. Proc. IUTAM Symposium on Unsteady Boundary Layers, Quebec, Canada, p. 570.

MCDONALD, H. (1978). Combustion modelling in two and three dimensions–some numerical considerations. Presented at Dept. of Energy Workshop on Modelling of Combustion in Practical Systems, Los Angeles. See also *Prog. Energy Comb. Sci.* **5**, 97 (1979).

MCDONALD, H. and FISH, R. W. (1973). Practical calculations of transitional boundary layers. *Int. J. Heat Mass Transf.* **16**, 1729.

MCGUIRK, J. J. and RODI, W. (1979). The calculation of three-dimensional turbulent free jets. *In* "Turbulent Shear Flows", **1**, 71. Springer. Verlag, Berlin.

MEIER, H. U. and KREPLIN, H. P. (1980). Experimental investigation of the boundary layer transition and separation on a body of revolution. *Z. Flugwiss.* **4**, 65.

MELLING, A. and WHITELAW, J. H. (1976). Turbulent flow in a rectangular duct. *J. Fluid Mech.* **78**, 289.

MICHEL, R. (1951). Etude de la transition sur les Profits d'Aile: Etablissement d'un Critére de Determination de Point de Transition et Calcul de la Trainée de Profile Incompressible, ONERA Dept 1/1578A.

MICHELFELDER, S. and LOWES, T. M. (1972). Report on the M-2 trials. International Flame Research Foundation, Doc. F36/2/4.

MILITZER, J., NICOLL, W. B. and ALPAY, S. A. (1977). Some observations of the numerical calculation of the recirculation region of twin parallel symmetric jet flow. Proc. Symposium on Turbulent Shear Flows, p. 18.11, Pennsylvania State University.

MONIN, A. S. and YAGLOM, A. M. (1971 & 1975). "Statistical Fluid Mechanics", vols 1 and 2. MIT press, Cambridge.

MOORE, J. (1973). A wake and an eddy in a rotating, radial-flow passage. *J. Engng. Pwr*, **95**, 205.

MOORE, J., MOORE, J. G. and JOHNSON, M. W. (1977). On three-dimensional flow in centrifugal impellers. ARC 37194.

MORSE, A. (1980). Axisymmetric free shear flows with and without swirl. Ph.D. Thesis, Imperial College, London University.

NAGUIB, A. S. and LOCKWOOD, F. C. (1976). Aspects of combustion modelling in engineering turbulent diffusion flames. *J. Inst. Fuel*, 218.

NAOT, D., SHAVIT, A. and WOLFSTEIN, M. (1973). Two-point correlation model and the redistribution of Reynolds stresses. *Physics Fluids*, **16**, 6.

NAOT, D., SHAVIT, A. and WOLFSHTEIN, M. (1974). Fully developed turbulent flow in a square channel. *Warme Stoffubertrag.*, **7**, 155.

NARASIMHA, R. and VISWANATH, P. R. (1975). Reverse transition at an expansion corner in supersonic flow. *AIAA J.* **13**, 693.

NAYFEH, A. H. (1980). Stability of three-dimensional boundary layers. *AIAA J.*, **18**, 406.

NEE, V. W. and KOVASZNAY (1969). The calculation of the incompressible turbulent boundary layer by a simple theory. *Phys. Fluids*, **12**, 473.

NIELSEN, P., RESTIVO, A. and WHITELAW, J. H. (1978). The velocity characteristics of ventilated rooms. *J. Fluids Engng.* **100**, 291.

PATANKAR, S. V. and SPALDING, D. B. (1970). "Heat and Mass Transfer in Boundary Layers". Intertext, London.

PATANKAR, S. V. and SPALDING, D. B. (1972). A calculation procedure for heat, mass and momentum transfer in three-dimensional parabolic flows. *Int. J. Heat Mass Transf.* **15**, 787.

PATANKAR, S. V., BASU, D. K. and ALPAY, S. A. (1977). Prediction of the three-dimensional velocity field of a deflected turbulent jet. *J. Fluids Engng*, **99**, 758.

PATEL, V. C. and LEE, Y. T. (1978). Calculation of thick boundary layer and near wake of bodies of revolution by a difference method. 12th Symposium on Naval Hydrodynamics, Washington, D.C.

PATEL, V. C. and NASH, J. F. (1975). Unsteady turbulent boundary layers with flow reversal. *Unsteady Aerodynamics* **1**, 191. Proc. Symposium held at the University of Arizona.

PATEL, V. C. and NAKAYAMA, A. (1976). Flow interaction near the tail of a body of revolution. *J. Fluids Engng.* **98**, 531.

PEACEMAN, D. W. and RACHFORD, H. H., Jr., (1965). The numerical solution of parabolic and elliptic differential equations. *J. Soc. ind. appl. Math.* **3**, 28–41.

PLETCHER, R. H. (1970). On a solution for turbulent boundary-layer flows with heat transfer, pressure gradients, and wall blowing or suction. Heat transf. 1970, Proc. 4th Int. Heat Transfer Conf., vol. 2. Elsevier, Amsterdam.

PLETCHER, R. H. (1978). Prediction of incompressible turbulent separating flow. *J. Fluid Engng.*, **100**, 427.

POPE, S. B. (1976). The probability approach to the modelling of turbulent reacting flows. *Combust. Flame*, **27**, 299.

POPE, S. B. (1977a). The calculation of separated boundary layers. Proc. Symposium Turbulent Shear Flows, p. 18.35. Pennsylvania State University.

POPE, S. B. (1977b). A rational method of determining probability distributions in turbulent reacting flows. To be published in *J. Non-Equilibrium Thermodynamics*.

POPE, S. B. (1979). A Monte Carlo method for the PDF equations of turbulent flow. MIT, Mech. Engng Dept. Report.

POPE, S. B. and WHITELAW, J. H. (1976). The calculation of near-wake flows. *J. Fluid Mech.* **73**, 9.

PRANDTL, L. (1945). Uber eine neue Formelsystem für die ausgebildete Turbulenz. Nachrichten der Akad. Wiss., Gottingen, Math. Phys., p. 16.

PRATAP, V. S. and SPALDING, D. B. (1975). Numerical computations of the flow in curved ducts. *Aeronaut. Q.* **26**, 219.

PRATT, D. T. (1978). Coalescence/dispersion modelling of high intensity combustion. 71st Annual Meeting of the A.I.Ch.E. See also AGARD CP 275, Paper 14, 1979.

PRIDDIN, C. H. (1975). The behavior of the turbulent boundary layer on curved porous walls. Ph.D. Thesis, Imperial College, London University.

PROUDMAN, I. and JOHNSON, K. (1962). Boundary-layer growth near a rear stagnation point. *J. Fluid Mech.* **56**, 161.

RADBILL, J. R. and MCCUE, G. A. (1970). "Quasilinearization and Nonlinear Problems in Fluid and Orbital Mechanics". American Elsevier, New York.

RAMSAY, J. W. and GOLDSTEIN, R. J. (1972). Interaction of a heated jet with a deflecting stream. NASA CR 72613.

RESTIVO, A. (1979). Turbulent flow in ventilated rooms. Ph.D. Thesis, Imperial College, London University.

REYHNER, T. A. and FLUGGE-LOTZ, I. (1968). The interaction of a shock wave with a laminar boundary layer. *Int. J. Non-Linear Mech.* **3**, 173.

REYNOLDS, W. C. (1976). Computation of turbulent flows. *Ann. Rev. Fluid. Mech.* **8**, 183.

RIBEIRO, M. M. and WHITELAW, J. H. (1980). The structure of turbulent jets. *Proc. R. Soc. London* **A370**, 281.

RICHARDS, B. E. (1976). Heat transfer to cooled turbine blades–a review. Von Karman Institute, Belgium. Lecture Series 83.

RICHTMYER, R. D. and MORTON, K. W. (1967). "Difference Methods for Initial-Value Problems". 2nd ed., Interscience Publishers, J. Wiley and Sons, New York.

ROACHE, P. J. (1972a). Finite-difference methods for the steady-state Navier-Stokes equations. SC-RR-72-0459, Sandia Laboratories, Albuquerque, New Mexico. (also Proc. 3rd International Conference on Numerical Methods in Fluid Dynamics, Paris, 3–7 July 1972).

ROACHE, P. J. (1972b). "Computational Fluid Dynamics". Hermosa Publishers, Albuquerque, New Mexico.

ROBERTS, D. W. and FORESTER, C. K. (1979). A parabolic computational procedure for flows in ducts with arbitrary cross section. *AIAA J.* **17**, 33.

RODI, W. (1972). The prediction of free turbulent boundary layers by use of a two-equation model of turbulence. Ph.D. Thesis, Imperial College, London University.

RODI, W. (1976). A new algebraic relation for calculating the Reynolds stresses. *Z. Angew. Math.* **56**, 219.

RODI, W. (1978). Turbulence models and their applications in hydraulics–a state of the art review–Univer. Karlsruhe SFB80/T/127; to be published by International Association for Hydraulic Research.

ROTTA, J. C. (1972a). "Turbulente Stromungen". Teubner, Stuttgart.

ROTTA, J. C. (1972b). Berechnung der kinetische Schwankungsenergie turbulenter Scherstromungen. DLR FB72-27, p. 347.

ROTTA, J. C. (1976). A family of turbulence models for three-dimensional thin shear layers. DFVLR IB-251-76 A25. Proc. 1st Symp. on Turbulent Shear Flows. Pennsylvania State University 1977.

ROTTA, J. C. (1979). Eine theoretische Untersuchung uber den Einfluss der Druckscherkorrelationen auf die Entwicklung dreidimensionaler turbulenter Grenzschichten, DFVLR-FB 79–05.

RUBESIN, M. W., *et al.* (1977). A critique of some recent second-order turbulence closure models for compressible boundary layers. AIAA Paper 77–218.

SAFFMAN, P. G. (1974). Model equations for turbulent shear flow. *Stud. appl. Math.* **53**, 17.

SAIY, M. (1974). Turbulent mixing of gas streams. Ph.D. Thesis, Imperial College, London University.

SAUL'YEV, V. K. (1964). "Integration of Equations of Parabolic Type by the Method of Nets" (Translated from Russian by G. J. Tee). Pergamon Press, The Macmillan Co., New York.

SCHNEIDER, G. R. (1977). Die Berechnung der turbulenten Grenzschicht am schiebenden Flügel unendlicher Spannweite mittels eines dreidimensionalen Mischungswegmodells. DFVLR-FB 77–73.

SCHUBAUER, G. B. (1939). Air flow in the boundary layer of an elliptic cylinder. NACA Report No. 652.

SCHUBAUER, G. B. and SKRAMSTAD, H. K. (1947). Laminar boundary-layer oscillations and stability of laminar flow. *J. Aeronaut. Sci.* **14**, 69.

SELCUK, N. and SIDDALL, R. G. (1976). Two flux spherical harmonic modelling of

two-dimensional radiative transfer in furnaces. *Int. J. Heat Mass Transf.* **19**, 313.
SFORZA, P. M., STEIGER, M. H. and TRENTACOSTE, N. (1966). Studies on three-dimensional viscous jets. *AIAA J.* **4**, 800.
SFEIR, A. A. (1975). The velocity and temperature fields of rectangular jets. *Int. J. Heat Mass Transf.* **19**, 1289.
SHABAKA, I. M. M. A. (1979). Turbulent flow in an idealized wing-body junction. Ph.D. Thesis, Aeronautics Department, Imperial College. To appear in *AIAA J.*
SHIR, C. C. (1973). A preliminary numerical study of atmospheric turbulent flow in the idealized planetary boundary layer. *J. atmos. Sci.* **30**, 1327.
SIMPSON, R., STRICKLAND, J. H. and BARR, P. W. (1974). Laser and hot film anemometer measurements in a separating turbulent boundary layer. Thermal and Fluid Sci. Center, S. Methodist University, Report WT3. See also *J. Fluid Mech.* **79**, 533, 1977.
SMITH, A. M. O. and GAMBERONI, N. (1956). Transition, pressure gradient, and stability theory. Proc. 9th Int. Cong. Appl. Mech., **4**, 234. Brussels, Belgium, (Also AD 125 559).
SMITS, A. J., YOUNG, S. T. B. and BRADSHAW, P. (1979). The effect of short regions of high surface curvature on turbulent boundary layers. *J. Fluid Mech.* **94**, 207.
SOCKOL, P. M. (1978). End wall boundary layer prediction for axial compressors. AIAA paper 78–1139.
SPALDING, D. B. (1970). Mixing and chemical reaction in steady confined turbulent flame. Proc. 13th International Symp. Combustion, Combustion Institute, USA, p. 649.
SPALDING, D. B. (1971). Concentration fluctuations in a round turbulent jet. *Chem. Engng. Sci.* **26**, 95.
SPALDING, D. B. (1972). The k-W model of turbulence. Imperial College, Mech. Engng Dept. Report HTS/72/12.
SQUIRE, H. B. and WINTER, K. G. (1951). The secondary flow in a cascade of airfoils in a nonuniform stream. *J. Aeronaut. Sci.* **18**, 271.
STEWARD, F. R., OSUWAN, S. and PICOT, J. J. C. (1972). Heat transfer measurements in a cylindrical test furnace. Proc. 14th International Symposium of Combustion, Combustion Institute, USA, p. 651.
STRATFORD, B. S. (1959). The prediction of separation of the turbulent boundary layer. *J. Fluid Mech.* **5**, 1.
SURUGUE, J. (Ed.) (1972). Boundary-layer effects in turbomachines. AGARDograph 164.
TATCHELL, D. G. (1975). Convective process in confined three-dimensional boundary layers. Ph.D. Thesis, Imperial College, London University.
TAYLOR, A. M. K. (1981). Confined isothermal and combusting flows behind axisymmetric baffles. Ph.D. Thesis, Imperial College, London University.
TAYLOR, A. M. K., WHITELAW, J. H. and YIANNESKIS, M. (1980). Measurements of laminar and turbulent flow in a curved duct with thin initial boundary layers. Imperial College, Mech. Eng. Dept. Report FS/80/29. Also NASA Contractors Report.
TELIONIS, D. P. and TSAHALIS, D.Th. (1973). Unsteady laminar separation over cylinders started impulsively from rest. Paper presented at 24th International Astronautical Conference, Baku, USSR.

TENNEKES, H. and LUMLEY, J. L. (1972). "A First Course in Turbulence". MIT Press, Cambridge, Mass.

THOMAN, D. C. and SZEWCZYK, A. A. (1966). Numerical solutions of time dependent two-dimensional flow of a viscous, incompressible fluid over stationary and rotating cylinders. University of Notre Dame Technical Report No. 66-14.

TOWNSEND, A. A. (1966). The flow in a turbulent boundary layer after a change in surface roughness. *J. Fluid Mech.* **26**, 255.

TOWNSEND, A. A. (1976). The Structure of Turbulent Shear Flow. University Press, Cambridge, England.

TRUELOVE, J. (1976). An evaluation of discrete ordinates approximation of radiative transfer in an absorbing, emitting and scattering planar medium. AERE Report R8478.

VAN DEN BERG, B., ELSENAAR, A., LINDHOUT, J. P. F. and WESSELING, P. (1975). Measurements in an incompressible three-dimensional turbulent boundary layer, under infinite swept wing conditions, and comparisons with theory. *J. Fluid Mech.* **70**, 127.

VAN DRIEST, E. R. (1956). On turbulent flow near a wall. *J. Aeronaut. Sci.* **23**, 1007.

VAN DYKE, M. D. (1975). "Perturbation Methods in Fluid Mechanics". Parabolic Press, Stanford.

VAN INGEN, J. L. (1956). A suggested semi-empirical method for the calculation of the boundary-layer transition region. Report No. V.T.H. 71, V.T.H. 74, Delft, Holland.

VISKANTA, R. (1966). Radiation transfer and interaction of convection with radiation heat transfer *In* "Advances in Heat Transfer" (Irvine, T. and Hartnett, J.P., eds), **3**, 175.

VKI (1979). Applications of numerical methods to flow calculations in turbomachines. Von Karman Institute Lecture Series 1979-7, United States NTIS reference N80-I2365.

WALDMAN, C. H. WILSON, R. P. and ENGLEMAN, V. S. (1974). Analysis of the kinetic mechanism of methane-air combustion and pollution formation. Report WSS/CJ 74-2 EPA-68-02-0270, USA.

WALKER, G. J. (1974). The unsteady nature of boundary-layer transition on an axial flow compressor blade. ASME paper 74-GT-135.

WALKER, G. J. and OLIVER, A. R. (1972). The effect of interaction between wakes from blade rows in an axial compressor on the noise generated by blade interaction. *J. Engng Power,* **94**, 241.

WALZ, A. (1969). Boundary layers of flow and temperature. Braun, Karlsruhe.

WANG, K. C. (1970). Three-dimensional boundary layer near the plane of symmetry of a spheroid at incidence. *J. Fluid Mech.* **43**, 187.

WANG, K. C. (1976). Separation of three-dimensional flow. Proc. Lockheed-Georgia Co. Viscous Flow Symposium.

WARD-SMITH, A. J. (1971). "Pressure Losses in Ducted Flows". Butterworth, London.

WAZZAN, A. R. OKAMURA, T. T. and SMITH, A. M. O. (1968). Spatial and temporal stability charts for the Falkner-Skan boundary-layer-profiles. Douglas Aircraft Co., Long Beach, California. Report DAC-67086.

WEINSTEIN, H. G., STONE, H. L. and KWAN, T. V. (1969). Iterative procedure for solution of systems of parabolic and elliptic equations in three dimensions. *I EC Fundamentals*, **8**, 281.

WHITACRE, R. and MCCANN, R. A. (1975). Comparison of methods of prediction of radiant heat flux distribution and temperature. ASME Paper 75-HT-9.

WILLIAMS, P. G. (1975). A reverse flow computation in the theory of self-induced separation. Lecture Notes in Physics, vol. 35. Proc. 4th International Conference Numer. Meth. Fluid Dynam. (Richtmyer, R. D., ed.) 445. Springer-Verlag.

WU, H. L. and FRICKLER, N. (1976). The characteristics of swirl-stabilized natural gas flames. Pt. 2: The behavior of swirling jet flames in a narrow cylindrical furnace. *J. Inst. Fuel,* **49**, 144.

WYNGAARD, J. C., ARYA, S. P. S. and COTÉ, O. R. (1974). Some aspects of the structure of convective planetary boundary layers. *J. atmos. Sci.* **31**, 747.

YAGLOM, A. M. (1979). Similarity laws for constant-pressure and pressure-gradient turbulent wall flows. *A. Rev. Fluid Mech.* **11**, 505.

Subject Index

A

ADI methods, 201
Advection (of turbulent kinetic energy), 3, 29, 62
Airfoil, 11, 41, 101, 110, 112, 219, 221
Algebraic stress model, 52, 53
Alternating direction implicit (ADI) methods, 201
Analogy, between heat and momentum transfer (Reynolds' analogy), 8–9; *see also* Prandtl number, turbulent
Annulus–wall boundary layer, 273, 275, 286, 289
Attachment-line (or stagnation-line) flow, 98, 257–258, 261–263
Averages, 22, 292–295
Axes, *see* Coordinate systems
Axisymmetric flow, 101, 108–109, 114, 152; *see also* Body of revolution

B

Bernoulli's equation, 93, 125
Biot–Savart law, 15
Blades, turbomachine, 273–276
  boundary layer on, 281–283, 285
Body force, 20; *see also* Buoyancy
Boundary conditions, 22, 34
  for thin shear layer, 94–98, 102, 106, 139, 160, 220
  for elliptic equations, 204, 206
  for Orr–Sommerfeld equations, 239
Boundary-layer approximation, *see* Thin-shear-layer approximation
Box method, 77, 88–92, 101–107, 123, 138, 159, 235, 264–266
Bubble, separation, 11, 123, 132–136
Buoyancy, 33, 73

C

"Cascade" of energy, 1, 15–16
Characteristic Box method, 152, 153, 266

Chemical reaction, with chemical control, 306–307
  equilibrium, 304–305
  with physical control, 299–305
Classification, of partial differential equations, 78–79
Closure problem, 5, *see also* Eddy viscosity, Mixing length, Modelling, Turbulence models
Coalescence–dispersion, 296
Combustion, 24, 290–309
"Complex" flow, 19
Compressible (variable-density) flow, 8–9, 24, 26, 74, 203, 224, 267
Computer program, 162–183, 186–188
Concentration, *see* Species
Conductive sublayer, 8–9, 13
Confined jet flow, 209
Conservation equations, 1, 22; *see also* names of conserved quantities
Continuity equation, *see* Mass, conservation of
Control-volume analysis, 2, 196
Convection (transport by mean flow), 23
Coordinate systems, 10, 101, 115–116, 151, 195, 204, 210–211, 254
Coriolis acceleration, 20
Corner flow, 21, 35, 210–215, 219, 228–234
Correlation, spatial, 6, 32
Crank–Nicholson method, 77, 85–87
Crossflow, 96, 138, 149–158
  profile formula, 96
Curvature, longitudinal, 20, 66, 75, 210–215
  transverse, 93, 108
Curvilinear coordinates, 151, 225
Cylinder, circular, in unsteady flow, 142–148

D

Density fluctuations, 8–9, 74, 292–295
Density-weighted averages, 292

Dependence, domain of, 78–79, 152–153; *see also* Upstream influence
Destruction, of Reynolds stress, 28
Diffusion (turbulent transport, especially of turbulent kinetic energy), 3–4, 28, 62, 66
Diffusion flames, 299–305
  molecular, 8–9, 25
Diffusivity, turbulent, 27, 73
Dilatation, 20
Disc, rotating, 247–252
Disc-stabilized flow, 207–208
Displacement
  effect, 219–223
  surface, 222
  thickness, 113, 114, 136, 148, 219
Dissipation of turbulent energy, 3–4, 6–7, 28, 38
  transport equation for, 6–7, 30–31, 45, 47, 51, 208, 294
Divergence or convergence, lateral, 20
Drag, 113–114, 221
"Drunkard's Walk" theorem, 15
Duct bend, 210–215, 232
Duct flows, 11, 21, 99, 123–131
  fully-developed, 124, 126
"DUIT" procedure, 134

E

Eddy, 15–16
  diffusivity, 27, 73
  viscosity, 6–7, 20, 21, 37–40, 93, 102, 110, 118–119, 127, 159, 169–170, 234, 256
  transport equations for, 37, 59
Eigenvalues, of Orr–Sommerfeld, equation, 235, 240–241, 243–245
Eigenvalue method, for duct flows, 128
Elliptic equations 34, 78–79, 192–194, 225
"e^n" method, 235, 248–252
Energy
  cascade, 1, 15–16
  equation, turbulent, 3–4, 28, 46, 62, 294
Ensemble averages, 22
Enthalpy, conservation equation for, 22, 24
Entrance region, of duct flow, 124–131
Equilibrium (self-preservation), 12
  local (energy), 12, 13, 38

Explicit numerical schemes, 77, 84
Extra strain rates, 1, 20, 69

F

"Fairly thin" shear layer, 10
Falkner–Skan transformation, 101, 108–109, 124, 160
  extension to three-dimensional flows 259–263
Finite-difference methods, 77, 101, 123, 139, 159, 192, 219
Finite-element methods, 80, 196
First-order closure, *see* Eddy viscosity
Flames, arbitrarily-fuelled, 308
  diffusion 299–305
  premixed 306–307
"FLARE" approximation, 134, 184
Free shear layers, 19, 233
  boundary conditions for, 95
Free-stream turbulence, 21, 273–274, 280
Furnaces, 309

G

Generation, of Reynolds stress, 28
Gradient transport, 46, 66–67; *see also* Eddy viscosity
Grid
  ("isotropic") turbulence 51, 71
  finite-difference 83–87, 140, 144, 152, 195–202, 264–266

H

Heat-conduction equation, 80, 84–92
Heat transfer, 8–9, 73, 290–291
  radiative, 24, 291, 296, 301
"Horseshoe" vortex, 15–16, 229, 286
Hyperbolic equations, 78–79

I

Impeller, *see* Rotor
Implicit numerical schemes, 84–92
Initial conditions, 96–98, 139, 161
  for three-dimensional flows, 152, 257–258

Subject Index

Injection, as simulation of displacement effect. 223
  for turbine blade cooling, 281–282
Inlet length, *see* Entrance region
Inner layer, 12–14; *see also* Logarithmic law, Sublayer
Instability, hydrodynamic, 17, 235–252
Integral calculation methods, 121
Integral scale, 6–7, 32
Interaction, between shear layers, 1, 11, 21, 124, 126–127
  viscous–inviscid, 114, 123–126, 219–227
  in three-dimensional flow, 269–271
Interface, *see* viscous superlayer
Intermittency, 43–44, 96–97
Internal flows, 21, 123–131, 207–215, 273–289
  boundary conditions for 95
Invariance, Galilean, 72
  rotational, 71
Inverse problems, for boundary layers, 99, 123–137, 159, 184–191
Isotropy, of small-scale motion, 15–16, 27
  return to 3–4

J

Jet, 11, 41
  confined, 209
  in crossflow, 216–217, 233

K

Kinetic energy, turbulent, *see* Energy

L

Large eddies, 6–7
Lateral convergence/divergence, 20
"Law of the wall", *see* Inner layer, Logarithmic law, Sublayer
Length scale, 5, 6–7, 32, 38–41, 69
  transport equation for 6–7, 12, 29, 32, 58, 68, 70; *see also* Dissipation
Linearization, of PDEs, 106–107
Local (energy)
  equilibrium, 12, 13, 38
  isotropy, 15–16, 27

Logarithmic law, 14, 50, 94, 106–207, 118–119
Low Reynolds-number effects, 13, 18, 43–44, 50–51

M

Mangler transformation, 101, 108–109
Mass, conservation of, 22, 23, 35, 78–79, 93
  transfer, 43–44; *see also* Injection, Transpiration
  -weighted averages, 8–9, 292
Mean value, *see* Averages
Mechul method, 128, 132–137, 185
Molecular diffusivity, 8–9
Molecular analogy with turbulent motion, 66–67
Momentum, conservation equation for, 2, 22, 23, 26, 93
  thickness, 113, 114
Morkovin's hypothesis, 8–9, 74

N

Navier-Stokes equations, 1, 23, 34, 78–79
  numerical solution of, 135, 193, 219
Net (finite-difference), *see* Grid
Newton's method, 104–105, 128, 186–187
Nose-region flow, 151–152
Numerical diffusion, 200
Numerical methods, 80, 159 and *passim*
  for elliptic equations, 201–205
  for time-dependent equations, 26, 138–148

O

One-equation model, 39–40
Orr–Sommerfeld equation, 235–252

P

Parabolic equations, 35, 78–79
Partial differential equations, 78–79
Partially parabolic equations, 213–215, 225

Passive scalar, 8–9, 27, 295; *see also* Species
Pipe flow, 11, 127
Poisson equation for pressure, 63–64
Prandtl number, 13
   turbulent, 27, 37, 46, 48, 54, 55, 73
Premixed flames, 306–307
Pressure, equation for, 63–64
   in thin shear layers, 1
   fluctuations, 3–4, 8–9, 53, 63
   gradient
      effect on boundary layer, 74
      effect on inner layer, 14
      normal, 225–227
   -strain "redistribution" term, 3–4, 28, 29, 58, 72
   /scalar-gradient term, 50, 56, 73
Pressure-corrector method, 201–202
Probability distribution, 295–298
   joint, 304–305
Production, of turbulent kinetic energy, 3–4, 38
Profile families, 96–97
"Pseudoviscosity", 200

R

Radiation, 24, 291, 296, 301
Rate of strain, 3–4
Recirculation, 34, 192–218
Redistribution, *see* Pressure strain term
Relaminarization or reverse transition, 13, 51
Reversed crossflow, 138, 149, 153, 155–158
Reversed flow, 132–136, 142–143, 147–148
Revolution, body of, 114, 149–158; *see also* Axisymmetric flow
Reynolds analogy, 8–9
Reynolds stress, 1–2
   transport equation for, 3–5, 28, 58–59, 70–72, 75, 225, 256
Rotation, about spanwise axis, 20
Rotor, 273–276
   tip gap effect, 288
Roughness, surface, 14

S

Scalar transport, 27, 33, 54–56, 73, 295; *see also* Heat transfer, Transport equations
Scales, of length, 12; *see also* Length scale of velocity, 12
Schmidt number, *see* Prandtl number
Secondary flow, skew-induced ("first kind"), 72, 213, 229–234
   stress-induced ("second kind"), 93, 98, 229, 232, 234; *see also* Slender flow
Self-preservation, 12
Semi-curvilinear coordinates, 10
Separated flow, 19, 99, 132–136, 147–148, 192–218, 267
Separation, 121, 146–148, 267
   bubble, 11, 123, 132–136
   in three-dimensional flow, 149–150, 152–158
Shear layer, 1, 10–11
   in "real life", 19
   interaction, *see* Interaction
Shock-wave/boundary-layer interaction, 224
Similarity, 108–109; *see also* Falkner-Skan transformation
Singularity, of boundary-layer equations, 132, 138
Skin-friction coefficient, 74, 111–114, 135–136, 146–148, 154, 157, 267
Slender shear layer, 10, 21, 219
"Source" terms, in transport equations, 24, 33, 38, 295
Species, conservation equations for, 22, 25, 54, 299–308
Spectrum, 26
Stability, hydrodynamic, 17, 235–252
   of numerical solutions, 84, 88, 138
Stagnation line flow, 98, 257–258, 261–263
Stator, 273–4
Strain, rate of, 3–4
Stream function, 102, 194
Stress, turbulent (Reynolds), 1–2
Stress-equation (transport-equation) model, 39–40, 71
Sublayer, conductivity, 8–9, 13
   viscous, 12, 13, 97
Superlayer, viscous, 18
Swirling flow, 209, 309

T

"TEACH" method, 203, 206–215
Thin-shear-layer approximation, 10, 35, 72, 93
Three-dimensional flow, 10, 72, 78–79, 235–241, 245–249, 251–272; *see also* Secondary flow
Time average, 22, 26
Trailing edges, 224, 254, 283
Transition, 1, 13, 17, 43–44, 51, 110–112, 235, 248–252
Transpiration, 14, 43–44, 281
Transport ("conservation") equations
    for dissipation, 6–7, 30–31, 45, 47, 51, 294
  elliptic flows, 192–195
  enthalpy, 22, 24
  length scales, 6–7, 12, 29, 32, 58, 68
  momentum, 2 and *passim*
  Reynolds stresses, 3–4, 21, 58–67
  scalar flux, 33, 295
  species, 27
  vorticity, 229
Transport terms, in Reynolds stress transport equations, 3–4, 66
  in length-scale transport equation, 12
Travelling-wave disturbance, 237
Tridiagonal matrix, 86–88, 91–92, 200
Turbomachinery, 273–289
  axial-flow, 273–276, 279–289
  centrifugal-flow, 273–274, 277–278
Turbulent (kinetic) energy, 3–4
  transport equation for, 3–4, 29, 42, 45, 51
Two-equation model ("k-ε" model), 39–40, 45, 204–218, 294

U

Unsteady flow, 98, 138–148, 159
Upstream influence, 35, 78–79, 144, 192, 219–220, 224
Upwind differences, 200

V

Van Driest mixing-length formula, 13, 43–44, 54
Variational equations, for duct flow, 129
Velocity-defect parameter ("wake parameter"), 18
Velocity-profile formula, 96–97
Viscosity, effect on turbulence, 13; *see also* Low Reynolds number
Viscous–inviscid interaction, 114, 123–126, 219–227
Viscous sublayer, *see* Sublayer
Vortex line, 15–16
Vortex stretching, 1, 15–16, 229
Vorticity, 1, 15–16, 194, 228–233

W

Wake, 11, 21, 41, 113, 115–120, 226
  parameter, 18
  turbomachine blade, 280, 283–285
Wall jet, 11, 42
  law of the, *see* Inner layer, Logarithmic law
Wing, 41, 113, 253–272
  -body junction, 10, 98, 229–231
  swept, 10, 253–272

Z

Zero-equation turbulence models, 39–40, 43–44; *see also* Eddy viscosity
Zig-zag Box scheme, 144–145, 153, 155–156

250,